The Violence of Representation

Essays in literature and society

Series editors: Nancy Armstrong and Leonard Tennenhouse

The Violence of Representation

Literature and the history of violence

Edited by

Nancy Armstrong

and

Leonard Tennenhouse

Routledge
London and New York

First published 1989
by Routledge
11 New Fetter Lane, London EC4P 4EE
29 West 35th Street, New York, NY 10001

Typeset by Mayhew Typesetting, Bristol
Printed in Great Britain
by T.J. Press (Padstow) Ltd, Padstow, Cornwall

British Library Cataloguing in Publication Data

Armstrong, Nancy
(Essays in literature and society)
The violence of representation: Literature and the history of violence.
1. English Literature, 1558–1980 — Critical Studies
870.9

Library of Congress Cataloging in Publication Data

The violence of representation.
(Essays in literature and society)
Bibliography: p.
Includes index.
1. Violence in literature. 2. Politics and literature.
3. Politics in literature. 4. Social conflict in literature. 5. Language and languages in literature.
6. Criticism – History – 20th century. I. Armstrong, Nancy. II. Tennenhouse, Leonard.
III. Series.
PN56.V53V56 1989 809′.93355 89–5881

ISBN 0–415–01447–6
0–415–01448–4 (pbk.)

In memory of Allon White

Contents

PART THREE CONTEMPORARY CULTURE: THE ART OF POLITICS

Introduction

Representing violence, or "how the west was won"

Nancy Armstrong and Leonard Tennenhouse

In the time that has passed since we began gathering these essays, the academic discourse around literary history and, more recently, cultural studies has changed remarkably both in Britain and in the States. Under the headings of new historicism, cultural materialism, socialist feminism, cultural poetics, and gender studies, work that deliberately violates traditional distinctions between poetics and politics has suddenly joined the mainstream of the humanities and social sciences. To document the change requires no more than a glance at newly revised departmental curricula, altered hiring priorities in a number of humanistic disciplines, not to mention the many announcements of new titles and new series to be published by the most prestigious university presses on both sides of the Atlantic. This is in addition to new journals in the humanities and social sciences featuring culture criticism. Even a cursory survey will show that programs not only in literature and history but also in cultural studies have taken up peaceful coexistence with the very tradition of research and scholarship to which they were once resolutely opposed.

What can it mean that such belligerent terms as "patriarchy," "resistance," and "subversion" are now cant phrases within literary criticism? Why is some mention of the class–race–gender nexus almost obligatory if one wants to appear up-to-date? For one thing, it means these topics have lost their oppositional edge. For criticism that once questioned the whole literary enterprise to have found a comfortable home within the humanities means that the literary criticism essentially hostile to it has performed some subtle but profound act of appropriation. Whatever it is that happens to cultural information when it becomes literature is happening with alarming reach and speed across vast territories of cultural space and time, thanks in part to the new interest in culture criticism and the institutional encouragement it has been receiving. In the name of doing non-canonical work, scholars throughout the humanities and social sciences have been extending literary critical methods into new areas which have never been read that way before. They are linking ideology to figuration, politics to aesthetics, and tropes of ambiguity and irony to instances of ambivalence and forms of political resistance. They have brought new understanding of the heterogeneity of any culture, including our own, and have clearly done much to revise traditional categories of period, style, and genre. But they have also scooped out these materials and incorporated them within contemporary academic discourse. We should be concerned, then, to ensure that our political self-consciousness keeps pace with the rapidly increasing use of certain interpretive procedures for reading other areas of our culture and cultures other than ours.

This collection necessarily makes a partial statement. All of our essays stay within the western European and pay special attention to the Anglo-American tradition. Our purpose is to set in place some of the components of our own history, namely, that of the discourse in which we are writing and of its ability to displace other cultural materials. As examples of how one may read the past as the history of the present, however, these essays should provide an idea of what literary history would look like if it were expanded to include the materials against which literature competed and won. By extending the techniques of reading figurative language, some essays demonstrate that seemingly artless political writing in fact clings fast to the poetics of high art. Others defy traditional cultural categories from a different angle, showing that writing which is usually regarded as highly aesthetic is no less politically engaged for being so.

While all our authors argue that politics and poetics are inseparable concerns, they locate violence at very different places within cultural production and, on this basis, can be roughly divided into two different camps. Some are interested in the symbolic practices through which one group achieves and others resist a certain form of domination at a given place or moment in time. These people are generally after an extra-discursive dimension of culture that will allow us to understand what is not ourselves. Others of us are less careful of the line between data and analysis, between forms of violence that are represented in writing and the violence committed through representation. They consider writing but one more symbolic practice among the others that make up cultural history. From their point of view, writing is not so much about violence as a form of violence in its own right.

But whichever side of the question concerning the relationship between violence and representation ultimately proves more convincing, our readers will no doubt expect such a volume as this to arrive at a definition of its key terms through one or more of a fairly limited number of avenues. Indeed, an early and obviously knowledgeable reviewer of this volume expressed dissatisfaction with our failure to do so. S/he wanted us to include an analysis grounded in the theories of Weber, Gramsci, and Althusser. We imagine the reader's reasoning went something like this: Does this volume intend to investigate collectively the changing relationship between historical subject and object worlds by analyzing the changing nature of cultural texts in mediating this relationship? Is it not in these terms, after all, that authors and readers alike must understand violence, its source, its instrument, its object, its effects, its historical significance?

In a word, no. We must confess it is not in these terms that we decided to pose the problem of this particular collection. To be sure, our readers will recognize the influence of Bakhtin and Foucault in the argument of our introduction as well as in the design of the collection as a whole. Even so, it is not from either of these theories of cultural history that we have abstracted the explanatory logic of our topic – that of the violence of representation – but from a well-known work of fiction by Charlotte Brontë. Deriving a definition of violence from *Jane Eyre* is our way of minding the lesson forcefully expressed in the essay by Teresa de Lauretis that concludes our history of violence. She claims that the discourse of theory, whatever its ideological bent, constitutes a form of violence in its own right in so far as it maintains a form of domination – "that of the male or male-sexed subject." That is to say, the only gender that can presume to speak as if ungendered and for all genders is the dominant gender.

Once aware of this discursive behavior, how do we deal with the reader's conviction that, when it comes to talking about something as serious as violence, only the theories of masculinist social science will do? We understand full well the authority that has accrued to such theory. It allows us to put in place distinctions between subject and object, personal and political life, text and theory, thereby establishing rational connections within and among the categories that constitute these oppositions. But theories capable of tearing away at this property of rational discourse from within – that is, theories of discourse – will not let us forget that the production of difference is never simply the production of differences such as these; with one and the same stroke, the production of difference also creates an opposition that subordinates one set of terms to another. The very oppositions we have mentioned indeed developed throughout the great imperial nations in the literature of government reports and the fledgling social sciences during the late eighteenth and nineteenth centuries. These oppositions subordinated objects and behaviors of the first (implicitly feminine) terms in each pair to those belonging to the second. In our own time, these oppositions mark not only the discourse of the professional-managerial classes but also that of the intellectuals who serve them. If it is the power of this discourse we want to name and historicize, then we can hardly exert such power without some willful attempt at calling attention to our own place within the field whose historical development we are tracing. Our use of *Jane Eyre* to define the violence of representation should be viewed as such a gesture.

We want to call attention to another inversion of the priorities our readers have every reason to expect from an introduction to a collection of essays on violence. Do not think for a moment that we are using Brontë's *Jane Eyre* because "she" allows us to view our discourse from a subordinated position and to speak on behalf of the racial, social, or sexual victim. On the contrary, we find Brontë's novel particularly useful because it exemplifies the other (feminine) half of liberal discourse. As such, it clearly demonstrates how such discourse, closely akin to literary criticism in this respect, suppresses all manner of differences by representing them as Otherness – that which one cannot be and still be part of the culture. Much like fiction that participates in the dominant discourse – as virtually all canonized or "literary" fiction does – literary feminism generally accedes to the terms of a rationalist, social science discourse that locates political power in men, in their labor, in the institutions they run, or else in certain forms of resistance to men or their institutions. Our reading of *Jane Eyre* suggests that while these constitute the acknowledged domain of political power, the site of its agency, the theater of its events, and thus the source of historical change, such power is not necessarily that which actually shapes people's lives in the novel; another source of power proves equally if not more compelling.

Although the world as Brontë represents it was once so governed by men and their money, that world dies out with the passing of Jane's uncle and Rochester's wife as well as with the amputation of his right hand and eye. In relation to these masculine modalities of power, Jane plays the role of triumphant underdog. Brontë has created a heroine conspicuously lacking in the acknowledged sources of social power – family (mostly dead), money (dispersed among surviving members), professional position (renounced), a pleasant disposition (only under the most conducive circumstances), and good looks (never much of a possibility). Nevertheless, as we will argue, Jane possesses something that allows her to exemplify and evaluate not only the thoughts and feelings of women but those of men and children as well. Jane can read, speak, and write. And by virtue of this power alone, she builds around herself a community that excludes those who do not think and feel and read and write as she does, all of whom die by novel's end. With Jane's ascendancy, the violence of an earlier political order maintained by overt forms of social control gives way to a more subtle kind of power that speaks with a mother's voice and works through the printed word upon mind and emotions rather than body and soul.

As American academics at this moment in history, we feel it is somehow dishonest to speak of power and violence as something that belongs to the police or the military, something that belongs to and is practiced by someone somewhere else. For clearly the subtler modalities of modern culture, usually classified as non-political, keep most of us in line, just as they designate specific "others" as the appropriate objects of violence. With a few notable exceptions, theory has yet to acknowledge adequately the forms of political power that operate in and through the modalities of personal life, care of the body, leisure-time activities, or literature, to name a few. And so we feel we must turn to *Jane Eyre* as a document that offers us a better chance to observe not only the power of normative culture but also its tendency to reclassify as "not political" such things as love, imagination, politeness, and virtue.

Writing depth: difference as Otherness

The first mode of violence in the novel is that which the novel represents. It is ''out there'' in the world on the other side of Jane's words. She encounters violence in the form of bad relatives, bad teachers, bad suitors, and, more generally, a bad class of people who have control over her life. These people are violent because their capacities of self are inferior to hers and because they suppress Jane's prodigious capacity for growth and development. That is to say, they are violent in a culture- and class-specific sense. In Jane's encounter with her aunt and surrogate mother, Mrs Reed, we bear witness to the first of several encounters between Jane's self and its Others:

> Mrs. Reed looked up from her work; her eye settled on mine, her fingers at the same time suspended their nimble movements.
>
> ''Go out of the room: return to the nursery,'' was her mandate. My look or something else must have struck her as offensive, for she spoke with extreme, though suppressed, irritation. I got up, I went to the door; I came back again: I walked to the window, across the room, then close up to her.
>
> *Speak* I must: I had been trodden on severely and *must* turn: but how? What strength had I to dart retaliation at my antagonist? I gathered my energies and launched them in this blunt sentence:—
>
> ''I am not deceitful: if I were, I should say I loved *you*; but I declare, I do not love you: I dislike you the worst of anybody in the world . . .
>
> . . . Shaking from head to foot, thrilled with ungovernable excitement, I continued:—
>
> ''I am glad you are no relation of mine: I will never call you aunt again as long as I live. I will never come to see you when I am grown up; and if any one asks me how I like you, and how you treated me, I will say the very thought of you makes me sick, and that you treated me with miserable cruelty.''
>
> ''How dare you affirm that, Jane Eyre?''
>
> ''How dare I, Mrs. Reed? How dare I? Because it is the *truth*. . . .''
>
> Ere I had finished this reply, my soul began to expand, to exult, with the strangest sense of freedom, of triumph, I ever felt. It seemed as if an invisible bond had burst, and that I had struggled out into unhoped-for liberty. Not without cause was this sentiment: Mrs. Reed looked frightened; her work had slipped from her knee; she was lifting up her hands, rocking herself to and fro, and even twisting her face as if she would cry.'' (36–7)

Though the first, this encounter provides the model for those shaping the rest of the novel, a novel whose magic is so powerful it still compels readers to make fast identification with its singularly unattractive heroine.

This heroine was created in the wake of bitter disappointment for Charlotte

Brontë. Her first novel *The Professor* was rejected for publication, while her sisters' novels – Ann's *Agnes Grey* and Emily's *Wuthering Heights* – were snapped up. Undaunted, Charlotte sat down and began the novel that would make her the most celebrated of the sisters, vowing (according to Mrs Gaskell's account) her heroine would outdo theirs by accomplishing everything they did without money, status, family, good looks, good fortune, or even a pleasant disposition. Thus we may assume Charlotte decided to outdo her sisters, or so she herself told Mrs Gaskell, by making something out of nothing at all – that is to say, making a self out of itself. In such a project, violence is an essential element. Each time Jane is confined to a room, kept at the bottom of a social hierarchy, silenced, humiliated, or otherwise crushed, we have more evidence that there is something already there to be confined, suppressed, silenced, or humiliated, something larger than its container, grander than any social role, more eloquent for all its honesty than those who presume to speak for it, and noble beyond their ken. This something is neither family, nor fortune, nor beauty, nor wit, nor accomplishments. For each act of suppression essentializes the subject – Jane herself: "Ere I had finished this reply, my soul began to expand, to exult, with the strangest sense of freedom, of triumph, I ever felt."

It is precisely this form of violence that Foucault identifies with "the repressive hypothesis" in *The History of Sexuality I*. Of the hidden "depths" that nineteenth-century authors discovered in the hearts of seemingly ordinary people, he has this to say:

> Whence a metamorphosis in literature: we have passed from a pleasure to be recounted and heard, centering on the heroic or marvelous narration of "trials" of bravery or sainthood, to a literature ordered according to the infinite task of extracting from the depths of oneself, in between the words, a truth which the very form of the confession holds out like a shimmering mirage. Whence too this new way of philosophizing: seeking the fundamental relation to the true, not simply in oneself – in some forgotten knowledge, or in a certain primal trace – but in the self-examination that yields, through a multitude of fleeting impressions, *the basic certainties of consciousness*. The obligation to confess is now relayed through so many different points, is so deeply ingrained in us, that *we no longer perceive it as the effect of a power that constrains us*; on the contrary, it seems to us that *truth, lodged in our most secret nature, "demands" only to surface; that if it fails to do so, this is because a constraint holds it in place, the violence of a power weighs it down*, and it can finally be articulated only at the price of a kind of liberation. (59–60, our italics)

In place of the repressive hypothesis, then, Foucault offers a productive hypothesis which inverts the logic of repression. To translate his argument into Brontë's terms, we would have to regard Jane's struggle against the forces constraining her as a discursive strategy for producing depths in the individual – what we have come to think of as the real Jane herself – that have been stifled in order for society to exist. The point of this discourse is to suggest there is always more there than discourse expresses, a self on the other side of words, bursting forth in words, only to find itself falsified and diminished because standardized and contained within the categories composing the aggregate of "society."

To understand the legacy of such a heroine and the magic she holds for readers today, Foucault implies, we have to identify a second modality of violence. Jane describes an obstacle in her path as a weight holding her down, if not in fact an act of violence against the self, and readers respond by rooting for her emancipation. So attached to the novel's heroine, we neglect to see how her descriptive power becomes a mode of violence in its own right. But in fact it does, as Jane reconstructs the universe around the polarities of Self and Other. We can observe this mode of violence at work, for example, in Jane's description of the fashionable Blanche Ingram, her rival for Rochester's affections:

> Miss Ingram was a mark *beneath* jealousy: she was *too inferior* to excite the feeling. Pardon the seeming paradox: I mean what I say. She was very *showy*, but she was *not genuine*: she had a *fine person*, many *brilliant attainments*; but her mind was *poor*, her heart *barren* by nature: *nothing bloomed* spontaneously on that soil; *no unforced natural fruit* delighted by its freshness. She was *not good*; she was *not original*: she used to repeat sounding phrases from books; she never offered, nor had, an opinion of her own. She advocated a *high tone of sentiment*; but she *did not know the sensations of sympathy and pity; tenderness and truth were not in her*. (188, our italics)

Surely no one can equal Brontë's narrator in the polite precision of such an evaluation. She deftly inverts Blanche's position of social superiority to Jane by employing an alternative system of value based on natural capacities of self ("she was too inferior to excite the feeling"). This move relocates value in depth ("mind" and "heart"). And in making this move, Brontë makes Blanche fall far short of Jane ("her mind poor" and "her heart barren"). Against the educational polish, the surface physical beauty, the "many brilliant attainments" of a Blanche Ingram, she thus pits Jane's ability to feel, which amounts to an ability to *be*.

It is important to notice just how aggressively the exertions of feeling on the part of the hapless orphan actually behave as they are put into words. We had a taste of this aggression in Jane's encounter with her aunt. Jane's ability to say what she feels was enough to drive her opponent, superior in every other respect, into speechless subjection. As she represents her rival for Rochester's affections, that person similarly appears to lack everything Jane possesses. Jane may not have any of the qualities society values, but, we feel, she is a real person. As a cultural object, too, she may seem to be much less perfect than Blanche. But the other woman's very accomplishments point to a deficiency of a more basic kind because hers is a failing of nature rather than culture ("nothing bloomed spontaneously on that soil"). Brontë uses nature on behalf of a class- and culture-specific embodiment of what Foucault, in the passage quoted above, calls "the basic certainties of consciousness."

For this self to become fact, it first had to dominate the different modes of identity present in the novel, just as Jane had to overcome certain Others in order to be a heroine. To earn the status of narrator, she must overcome Blanche, Mrs Reed, Mr Brocklehurst, virtually everyone and anyone who stands in her way. This is the violence of the productive hypothesis: the violence of representation. To be sure, every mode of identity contending with Jane's identity as a self-produced self poses a threat to that self. But in order for her to emerge as the knowledgeable spokesperson of other identities, these differences must be there and reveal themselves as a lack, just as Blanche ceases to be another person and become a non-person. The same process that creates Jane's "self" positions "others" in a negative relationship to that self. The violence of representation is the suppression of difference.

Until the very end of the novel, Jane is always excluded from every available form of social power. Her survival seems to depend on renouncing what power might come to her as teacher, mistress, cousin, heiress, or missionary's wife. She repeatedly flees from such forms of inclusion in the field of power, as if her status as an exemplary subject, like her authority as narrator, depends entirely on her claim to a kind of truth which can only be made from a position of powerlessness. By creating such an unlovely heroine and subjecting her to one form of harassment after another, Brontë demonstrates the power of words alone. Indeed, she demonstrates that words are all the more powerful for being alone. Only when they appear to come from a position outside the field of power do they appear to speak for everyone. If Jane is always the victim, one tends to assume her power is nothing other than that of pure goodness and truth. Indeed, how many of us have felt the novel's hegemonic effect without noticing the power of representation as Jane unleashes it on the various institutions she encounters on her adventures? Yet it is clear that in this novel one must possess depth or die. The only ones who survive at the end of Jane's story are those who resemble Jane or someone she could marry or one of her offspring. She is the progenitrix of a new gender, class, and race of selves in relation to whom all others are deficient. No less is at stake in granting such an individual power to author her own history.

Herein too lies something like the historical beginnings of modern interpretative procedures. When Charlotte Brontë, herself steeped in the science of phrenology, had Jane infer depth (in Blanche's case, a lack of depth) from the body's surface, the author effectively detached consciousness from a point of origin within the physical body. Brontë was one of a number of writers, intellectuals, and professionals who represented a self-originating body of consciousness, separate from, though housed within, the physical body. They rewrote the body and showed how it could be read, in turn, as revealing the particularities of consciousness. It may seem a small thing, the slippage that produced a source of the self that is within but not of the body, a minor event in cultural history perhaps, but what a difference it made!

Writing culture: how the representation of violence became the violence of representation

We have offered a crude distinction between two modalities of violence: that which is "out there" in the world, as opposed to that which is exercised through words upon things in the world, often by attributing violence to them. But our ultimate goal is to demonstrate that the two cannot in fact be distinguished, at least not in writing. We used *Jane Eyre* to demonstrate that violent events are not simply so but are called violent because they bring together different concepts of social order. To regard certain practices as violent is never to see them just as they are. It is always to take up a position for or against them (as when Jane expresses ingratitude for the treatment she receives in the Reed household.)

We have used Brontë's novel to explain what we mean by the violence of representation because this violence appears there in its most benign, defensive, and nearly invisible form – a power one can use without even calling it such. Like *Jane Eyre*, the narrative to follow is neither theory nor an arrangement of information that adheres to and validates theory but a sequence of readings arranged so as to challenge rationalist representations of the nature, cause, and effect of cultural change, representations that suppress the emergence of their own discursive power where fiction reveals it. In introducing the essays included in this volume, we hope to show the unfolding of a paradigm: the development of sophisticated technologies of the individual and its Others. Successive phases of imperialism – extending down through other classes within the imperial nations as well as outward to the colonies and informing foreign policy to this day – have turned the violence of representation into the ubiquitous form of power that is the ultimate though elusive topic and target of this book.

We had another reason for using *Jane Eyre* to exemplify the paradigm of violence that will be developed, qualified, and finally transformed in the essays that follow: the novel was written by and spoken through a woman. In distinguishing two modes of violence and then linking them together, we have suggested something of the centrality of gender (and thus of women as the gender-marked bodies) in any discussion of power. When we showed that Jane's power depends to a great extent on the fact that she is outside the field of competing economic and political interests, furthermore, we isolated a problem haunting any collection such as this. Our own model of culture is implicated in the very form of power it sets about to critique: a model of culture that is constructed around the viewpoint of a specific gender, class, and race of people. Like Jane, we tend to think of ourselves as outside the field of power, or at least we write about "it" as if it were "out there." That is to say, we situate ourselves in a "female" position relative to the discourses of law, finance, technology, and political policy. From such a position, one may presume to speak both as one of those excluded from the dominant discourse and for those so excluded. But doing so, we would argue, is no more legitimate than Jane Eyre's claim to victim status. Within liberal discourse, what might be called the male and female positions are represented as if they could contain all other differences. By implication, then, we will be tracing the history of our own authority along with that of the modern subject.

Early modern culture: putting the politics back into poetics

It is the relationship between the traditional arena of political events and those domains of cultural history considered merely contingent that Stephanie Jed explores in her study of fifteenth-century Italian humanism. She contends that despite the tradition of representation that sets humanist learning apart from the political tyranny with which this learning coexisted, literature is implicated in politics from the beginning. What Jed calls the humanist code contains a message of political violence both in tales of rape and murder and in the metaphors that displace this violence. More importantly, argues Jed, the formation of the tradition that grew up around this code constituted an attempt on the part of humanists "to dominate the scene of writing with exclusively 'literary' models." It is their practice, she argues, to privilege "classical learning as the true source of political understanding and aspirations for freedom." This is certainly how we understand the past, and it was just as certainly how humanists then sought to be understood in relation to those, including Leonardo da Vinci, who did not happen to possess the privileged form of literacy although he did possess other forms. These "other" forms of writing record many of the events we consider history to be made of, yet they are not considered fit objects of literary study. And they do not contribute therefore to our understanding of who we are as purveyors of specialized literacy in an age of mass culture.

Thus Jed implies a continuity to the violence of representation when she describes fifteenth-century humanism as the beginning of "mass literacy." This tradition of writing was tied to political discourse and conveyed the message of domination even as it distanced itself as "literature" from politics and concealed that message in natural metaphors that legitimated violence on the part of a nation and class. In Jed's view, the consequences of unselfconsciously accepting the humanistic distinction between literature and politics is grave:

> by concealing the social specificity of the humanistic code to a particular kind of "literacy" or cultural practice, humanism obstructs the access of other written codes to representation by the sign "literature." For example, neither the "literature" of fascist Italy nor fifteenth and sixteenth century archival documents have yet, for the most part, made their way . . . into our courses on that category of writing – "literature" – which claims to transcend our everyday concerns. And we find it difficult to connect the narrative of the violence of tyrants codified by humanism as "literature" to the violence represented daily in court records, in documents emanating from the Pentagon, or in a communique issued by the death squads in El Salvador.

Modern humanism suppresses the political content of metaphors, then, as well as alternative literacies that would allow alternative political viewpoints to exist. By repeating the behavior of an earlier humanism, the modern institution of "literature" does much the same thing. It keeps us from focusing on the worst abuses of power on the part of our government; indeed the chairman of the National Endowment for the Humanities, Lynn Cheney, declared in 1988 that political matters are not supposed to be our concern.

Peter Stallybrass's essay deals with English culture at about the same period of time, but one would be hard put to place his work in the same discipline with Jed's, much less in the same epoch. Stallybrass clearly announces his perspective as that of the "bottom-up" cultural historian. Using the figure of Robin Hood to argue for the variety of ways in which charivari was made and remade as a contested domain of social practices, Stallybrass claims for it and like practices (for which, from Bakhtin, he borrows the term "carnivalesque") a subversive potential. Charivari does violence to such social orthodoxies as the rites of the Church (we are treated to a cuckolded bishop riding backwards on a horse) and displays of state power (as when an army is turned back by the bare-assed boys of Norwich). In time, however, the violence of carnival is successfully included in Church ritual and appropriated by the state, marking the close of the early modern period and the beginning of an age when such performances of violence were displaced by rhetoric. By the Restoration, as Stallybrass conveys the irony of a double inversion, Robin Hood himself was used to acknowledge the authority of the crown: "In this way, political and social hierarchy could be effaced by the elision of monarchy, play, and carnival in the ideological construction of England as a band of brothers."

Thus by beginning from a different array of materials than Jed's, Stallybrass projects a very different cultural history. He is intent on showing how practices other than writing, representing innovative responses on the part of local groups, resisted the procedures of land enclosure and forest restriction and thus provide us with a contestory underside to the history of dominance. In such a history, the stress is on the creative side of violence:

the inversion of hierarchical structures, subversion of official policies by local practices, and the successful violation of law. Violence here becomes the sign of conflict between two different concepts of justice – one that guaranteed even poor men certain rights, and one based on the rights of property: "The relation of 'play' to 'law' could be constituted quite differently by the subordinated classes. In carnivalesque 'play,' the law became the subject of interrogation. Indeed official law could be seen from this perspective as a form of *disorder*."

Both Jed and Stallybrass organize analysis around the same oppositional qualities of discourse: dominant vs. subordinate, national vs. local, written vs. oral. Both attempt to show that a particular order survives because it is dominant. Each acknowledges residual or contestory practices that failed to find a place in modern history because they were deemed disreputable. Thus they implicitly establish a continuity between the fifteenth century (Jed's "humanism" and Stallybrass's "rights of property") and our own day, even though, for Stallybrass, the connection is not one made through our practices of reading and writing and, for Jed, it is. Leonard Tennenhouse's view of early modern England seems to question the fact of such continuity, arguing that it is more a product of modern humanism than one of fifteenth and sixteenth century modes of domination.

Neither our world nor our selves, in his estimation, are made of the same stuff that Shakespeare's were. Despite modern efforts to make his heroes and heroines resemble those formulated in romantic poetry and realistic fiction, our modes of characterization do not come from drama that was performed at the favor of the monarch. Rather, our idea of who people are and how they think and behave grew out of an oppositional tradition that emerged during the modern period and whose hegemony was assured by the massive rewriting of Renaissance drama that occurred during the Restoration period. Tennenhouse chooses an inflammatory topic – the violent treatment of aristocratic women's bodies on the Jacobean stage – to insist on this difference. Thus his essay has several different notions of violence going at once: one depicted on the Renaissance stage, one enacted on a popular audience through such displays of power, and one effected through our appropriation of this past in the name of modern bourgeois humanism. This last, according to Tennenhouse, falsifies the place of public theater in earlier societies. Contemporary humanism translates the materials of an earlier social body into those of our own gendered body, allowing us to infuse the figures of an earlier epoch with desires and needs characteristic of our own. Of the brutal treatment of aristocratic women that seems to have been an obligatory element in Jacobean tragedy, he explains,

> Tragedies . . . stripped away the very qualities that had distinguished heroines of a few decades before. On the Jacobean stage, we see willful aristocratic women punished for possessing the very features that

> empowered such characters in Elizabethan romantic comedy. The ritual purification of their bodies did not simply give vent to misogynistic impulses (although I am sure that it did that as well, and indeed continues to do so), it also revised the political iconography of an earlier monarchy, which was understood by English monarchs to be a very real instrument of power. Only this could have made the assault on the Elizabethan style of female so pervasive.

By draining the politics out of Renaissance poetics, in Tennenhouse's analysis, criticism tends to do much the same thing as humanism in Jed's analysis: suppress the role literature has played in political history.

Not unlike the scaffold, as he describes it, the public theater produced an image of society divided into two bodies – the aristocratic body, or body of blood, and the mass body which represented that of the populace. And like the scaffold, the theater placed the two in a radically disymmetrical relation of power. While Elizabeth occupied the throne, any threat to the body of an aristocratic female, as illustrated by Shakespeare's Lavinia, was nothing less than an assault on the nation itself. With the aging of Elizabeth, however, the theater ceased to identify the female body with the aristocratic body and, in this way, with the fate of the nation. As symbolic sites of pollution, women were subjected to ritual purification on the stage.

In the world Tennenhouse describes, it would be impossible to think of gender and class as separate formations as we tend to do in the modern period. But while gender is always used to represent class in Renaissance culture, it is not the difference that matters most, as it does in ours. What matters more is the fate of the aristocratic body, the difference between it and what Bakhtin calls the mass body. Tennenhouse suggests that, at least in Renaissance drama, there is no such thing as gender autonomy, or woman *per se*. There are only women who belong to the aristocratic body and those who do not. Gender carried with it entirely different symbolic properties depending upon its class designation, and destiny depended first on one's relation to the elite community and secondarily on one's relation to men and other women.

What becomes visible as we draw together these bits of early modern culture is a highly developed rhetoric of violence that was no less rhetorical for being played out on the scaffold and (parodically) enacted in carnivalesque inversions such as charivari, (analogically) performed in the ritual torture of women on the stage, and (figuratively) rendered in the metaphors and allusions of early humanistic discourse. However disparate they seem to us now, these practices may, for all we know, have played comparable and even, in some respects, interlocking parts in a single process whereby political identities were produced, positioned in relation to one another, contested, negotiated, and inevitably revised. Given the form in which these materials have come down to us, one thing we can say with some certainty is that their relationship to one another and to the practices with which they competed for meaning was lost as they passed into literature. Restoring the political dimension of materials that have been taken into the literary canon or excluded from it involves authors in another web of competing interests, a contemporary battle for representation.

Appearing to echo other authors, George Mariscal would situate the literary work in dialogue with other works and other practices contemporaneous with it. To do this for a masterwork of Spanish literature, however, he finds he must urge a correction of the Foucauldian view (of Tennenhouse), itself a correction of the history from below (offered by Stallybrass). In the opposition that emerges when ''the two Quixotes'' are set in historical dialogue, Mariscal determines that the ''kind of spectacle'' represented by Avellaneda's narrative ''has less to do with the display of the 'unrestrained presence of the sovereign' (as Foucault would have it) than with the complicity and participation of the entire social body in the containment of anomalous subjectivity.'' Cervantes' Quixote is brought back in line by another Quixote, who does not, as it turns out, make it into the literary tradition. Over the course of time, Cervantes' Quixote turns out to be the true Quixote because he gives voice to modern subjectivity – willful, potentially subversive, at odds with the state – in tentative proto-modern form. By detaching the Cervantine figure from its historical double, Mariscal implies, we postpone the day of considering the subject as an historically produced and politically contested actor in the ongoing drama of hegemony.

By suppressing the ''other'' Quixote, our literary tradition does not allow us to see how the Cervantine subject was subordinated by more powerful groups within his own aristocratic class. The Anglo-American tradition in particular tends to represent the ''true'' Quixote as evidence of the triumphant individualism which is of a piece with modern humanism:

> In this new historical conjuncture, what was perceived as the ''rebellious'' subjectivity of the 1605 Quixote towered over the aristocratic and traditionalist ideologies which informed the texts of both Avellaneda and Lesage and made them undesirable. Avellaneda's attempt to punish Cervantes had finally failed, at least in the English-speaking world. What had always been two Quixotes had now become one.

In fact, Mariscal contends, the two took shape in two very different political arguments, offering in Cervantes' day a critique of aristocratic domination from within, a minority perspective, and, in our day, testimony to the universality of a certain notion of power through the celebration of individual subjectivity.

We emphasize this attempt to rescue the bourgeois subject – and throughout the essay Mariscal's defensiveness on behalf of that subject – because it both sets him apart from and establishes the nature of his affiliation with the contributors we have already discussed. Violence well in the past has a certain legitimacy for all of them because they believe it was honestly political then. They speculate that everyone knew what it meant to commit certain acts of violence and that a sufficient number of the people who mattered felt violence served the interests of the entire social body. Yet it is safe to say that none of the authors in this volume would themselves welcome a political order which subjected individuals' bodies to the will of a monarch. What do they have at stake, then, in contesting the nature of the power represented in Renaissance literature? Clearly the form that power takes for good or ill has everything to do with power now, what we take it to be, and how we are to position ourselves in relation to it. And, in taking a position on the past, each of our contributors inevitably takes up a position in the contemporary struggle for representation. Each challenges some of the assumptions that might have allowed us to read other cultures through the modern humanistic grid.

Stallybrass privileges the perspective created through ritual inversion not as that of the authentic culture of the folk, we should add, but as a perspective enabling one to view domination from the position of the dominated. Jed and Tennenhouse consider the ways in which practices that eventually made their ways into the literary canon actually authorized what we now take to be intolerable forms of political power. This is in the face of modern humanism's tendency to see earlier humanist writing in opposition to the political tyranny of its day. Mariscal takes on the whole issue of bourgeois subjectivity. He reads it in relation to a dominant aristocratic mentality characteristic of seventeenth-century Spain, on the one hand, and in contrast with the kind of individualism that dominates modern culture, on the other. Together, then, these essays ask us to see modern culture as the hegemony of a new form of subjectivity which tended to suppress the possibility of other subject positions as well as its own historicity.

Modern culture: the triumph of depth

With the European Enlightenment, it becomes impossible to consider the representation of violence separate and apart from the violence of representation. We move from a situation where power is represented as being openly and avowedly violent to one in which the state becomes more illusive and allusive at once. Reference is always being made to it, according to those of us who work in this period, but it is never really there as an embodied presence. The power with which this group of essays concern themselves is that of hegemony itself – the capability of certain cultural formations to position everything else in a negative relationship to it.

In J. M. Coetzee's study of early travel writing about South Africa, we see this dynamic most clearly "at work" in a kind of writing that established the cultural ground justifying European imperialism until very recently. Underlying early narrative accounts of the Cape one finds the same categories (for instance, dress, diet, housing, medicine, government, recreation, economy, character) that would shape, not only anthropology, but also the massive world of discourse that developed around productive labor in order to exalt those who worked. As Coetzee tells the story, this began with the Reformation in Europe but quickly turned into the means of waging war on "social parasitism." Early descriptions of Hottentot culture – what Coetzee calls "the discourse of the Cape" – provide him with a way of isolating this phenomenon where its onset may be clearly perceived. And the war for representation continues in these terms. What Coetzee says about the operations of this writing reverberates down through the centuries and holds true for all the great imperial nations:

> The anathema on idleness . . . did not falter with the Enlightenment; for the revolt against authority that constituted the Enlightenment simply replaced the old condemnation of idleness as disobedience to God with an emphasis on work as a duty owed by man to himself and his neighbour. Through work man embarked on a voyage of exploration whose ultimate goal is the discovery of man; through work man becomes master of the world; through a community of work society comes into being.

The Lockean notion of culture as something at work is still deeply imbedded in our culture and ourselves. Indeed, it is shaping our writing – our "work" – even now.

Peculiar to this discourse at its inception, in Coetzee's account, and one reason why it is an effective instrument of imperialism, is the compulsion to represent what is culturally Other as a negation of Self. So represented, we cannot help but take otherness personally. Thus the idleness of the Hottentot comes to be seen as an act of violence. Consider, for example, in a quote from an early moment in the discourse of the Cape, how the passivity of the indigenous culture takes on the violent overtones of cannibalism, contagious disease, and promiscuity, wherever that culture violates European categories:

> Their food consists of herbs, cattle, wild animals and fish. The animals are eaten together with their internal organs. Having been shaken out a little, the intestines are not wasted, but as soon as the animal has been slaughtered or discovered, these are eaten raw, skin and all . . . . They all smell fiercely, as can be noticed at a distance of more than twelve feet against the wind, and they also give the appearance of never having washed.

Randall McGowen's essay picks up this same thread in the legal discourse of early-nineteenth-century England.

To begin his investigation of the process which made violence the provenance of the Other, McGowen declares: "There was a time when violence was not considered the central ingredient of crime, a time when the state was legitimated by its violence, a time when violent acts were attributed to God's righteousness. Our construction of and preoccupation with violence and its consequences are of recent origin.'' McGowen's research in legal history covers the first three decades of the nineteenth century, the same period when the modern middle classes were solidifying their hold over British culture. His point is to articulate an all-important relationship between humanitarian claims made on behalf of the individual and the attribution of violence to others. Legal discourse was one means of displacing the capacity to perpetrate violence upon members of the social body from the ''head'' of the state onto its lowest or most marginal members, if not onto those who had no claim to membership in society at all. From this simple but pervasive shift of the moral weight in legal argument, a double consequence ensued. As government positioned itself as the defender of individuals over and against those who would do violence to them, crimes became crimes against persons and property rather than crimes against the state, and those who did not have something to gain or lose by adhering to property laws were implicitly criminalized.

We would not want to claim (nor, we are sure, would McGowen) that such a change in the nature of the subject and his relation to the state took place at this moment or at one discursive site alone. Rather, it is in a legal discourse of the early nineteenth century that a far more pervasive change found perhaps its clearest articulation. Claims made on behalf of men who worked had by then produced a notion of the individual – autonomous, ambitious, and responsible for himself – throughout the western European nation states. Ideally a rational and independent male, this individual transformed the idea of the state and forged new national identities. Born in representation (what Raymond Williams calls ''the long revolution''), the modern British subject did not develop on analogical planes of literature and life, mind and society. And as Foucault's two epic accounts of this process (*Discipline and Punish* and *The History of Sexuality*) throughout Enlightenment Europe illustrates, the modern subject had many transformations and gave its shape to those who used its metaphors (of enclosure, internal division, and development) as well as to the institutions that these individuals formed in order to socialize others. At some point, we imagine, this individual – at first largely a rhetorical figure in whose name specific charges were leveled against the state and specific rights claimed – began to take root in the heads of certain people, as the figure of what they were or what they desired to become. Thus the modern individual passed from rhetoric into the domain of nature. Rather than claim rights in his name over and against a state considered insensitive to individual needs, at some point the new literate classes grew defensive. They ceased to think of themselves in opposition to an older landowning class. They began to speak for the interests of property owners in the face of opposition from an organizing proletariat. E.P. Thompson's magnificent account of *The Making of the English Working Class* identifies this repositioning of the modern middle-class subject as one that occurred during the same time period as the legal debates over punishment.

It is as part of this larger history of the modern subject that we think McGowen's argument becomes especially telling. In his words,

> The discourse upon violence produced a new way of reading society. It generated a different set of responses to crime along with a different constitution of criminality from that of an earlier code. These images of crime demanded new remedies that in turn projected a different face of government.

This, we would like to suggest, marked the transformation of the master narrative of British culture from one about an individual nobly contending against an obstructive and tradition-bound state into one about a paternalistic state caring enough to rehabilitate individuals who were deeply flawed. *Jane Eyre* offers a clear illustration of this transformation. The process by which she develops into a deep and internally conflicted self capable of overseeing the development of others is also a process that revises (indeed, feminizes) the image of society. The social institutions that suppress her become those over which she is in charge – where power consequently assumes the benevolent forms of supervision and education instead of force. As we have argued at length in our discussion of Brontë's novel, the relocation of violence within individuals, its presence in whoever designates them as Other, marks a shift in the form of political power from violence to representation.

The gradual disappearance of the scaffold and pillory was but one in a vast constellation of changes that separated people according to the categories of Henry Mayhew's massive ethnography *London Labour and London Poor* into those who worked and those who did not. Especially subject to assault in such studies as likely to be in violation of these categories were women, aborigines, vagabonds, and riff-raff of various kinds, and a great deal of energy was devoted to insisting this was so. Allon White's study of "the end of carnival" traces a basic change in the organization of culture itself across the social map of western Europe. In his view, the modernization of Europe brought about not only the gradual suppression of a communal festive tradition culminating in its isolation at the margins of towns and continents and the "dramatic collapse of the ritual calendar" but also the "displacements" of this tradition onto "bourgeois discourses, like art and psychoanalysis." White uses the discourse proliferating around hysteria in the closing decades of the nineteenth century to demonstrate the degree to which "the repudiation of reveling in mess, which was a central feature of carnival, was a fundamental aspect of bourgeois cultural identity." In so doing, he brings the process we have been tracing full circle.

Coetzee's study of the discourse of the Cape identifies the beginning of modernity with the production of a discourse representing culture itself with the idea of work. Idle people are outside of culture. Where otherness as idleness has violent overtones from the beginning, it waited for the late eighteenth and early nineteenth centuries to actually remodel British common law around this principle. It assumed those who did not possess property, knowledge, or skills posed a potential threat to those who did. Implicitly these reforms, those carried out through the regulations concerning burial practices, and many others as well, conspired with the banishment of carnival and the institution of the work week to criminalize those who would not or could not comply. To borrow a phrase from Raymond Williams, what might be called a new "structure of feeling" was born. The ethnographic material which in centuries past had enabled people of diverse regions, classes, political factions, and sects to feel part of a social body were now associated with marginality, the world of night, and a loss of social identity (which had come to be represented exclusively in the terms of respectable culture).

White describes something like an internalization of that material to represent what one could not be and still be oneself. Divided into a purified legitimate body of people and the criminal underworld of loafers, drunks, and prostitutes, the state had become a state of mind. We might imagine a sexual superego developing in response to the hegemony of respectable culture and its denigration of what was represented as the disgusting lower parts of the old social body. At least this is how White poses the social-historical problem raised by the outbreak of hysteria throughout the great imperial nations of western Europe:

> What happens when a hegemonic group destroys the physical ritual practices of a whole society and then endeavours to utilize the symbolism and purely discursive forms of those rituals for its own end? To put it more pointedly, part of the very process of moving from a rich, physical culture of the social body through to textual representation has been necessarily a repression of the body through the "agencies of disgust."

Hysteria and related impairments of the sensual body were one sign "the carnival was over" and a "strange carnivalesque diaspora had just begun."

By examining another domain considered off-limits to the state, Vassilis Lambropoulos provides us with yet another account of the process by which the power of the modern state scattered and spread to become something else, something even more powerful. Still thought by many to be beyond the reach of culture itself was the fantasy life of respectable young women. Another consolidation of middle-class power took place with the establishment of the sacred preserve of high culture:

> By the mid-nineteenth century, the idea of secular salvation had taken many forms, including those of progress, scientific knowledge, racial purification, class revolt, and Puritanism. But all these forms were too immediate, too tangible to inflame and keep alight popular imagination, not

> to mention the mystical needs of the intellectual sensibility. A grandiose scheme, larger than life and purer than history, was required for the depiction of the present and the revelation of the future.

Lambropoulos emphasizes the degree to which Arnold's idea of culture requires that of an impending anarchy: "Under the emergency created by the imminent danger of anarchy, the dissolution of order and the collapse of authority, the defense of authority becomes the absolute priority."

Lambropoulos has chosen this particular essay of Arnold's presumably because it so clearly elides the political and the aesthetic through the term "anarchy." In the political lexicon, according to Arnold, "anarchy" is the result of a longstanding claim to freedom on behalf of all Englishmen. While Arnold is careful never to contest the legitimacy of this claim, he does quarrel with the way they (meaning lower- and middle-class Englishmen) tend to put it into practice. As Lambropoulos notes, "they" tend to "march," "meet," "enter," "hoot," "threaten," and "smash." Freedom, when put into practice, threatens social order; it is violent. Thus Arnold locates the source of violence "in the sovereign subject, that is, the individual of the modern, post-Revolutionary era." In his model of culture, then, we observe something remarkably like the conditions that, according to White, manifest themselves in *fin-de-siècle* hysteria.

An older conception of culture that respected the autonomy of the individual (so long as he worked, of course) now poses a threat. Arnold sees heterogeneity as violence. This is where the aesthetic, or bourgeois concept of "high" culture, comes into play: "when people become acculturated, they are simultaneously subjected to rational control. Culture gives authority the ultimate justification – inherent value; in addition, by training subjects, it makes coercive subjection redundant." The problems of culture and neurosis are one and the same. Or rather, to solve the political problem of recentralizing British authority, Arnold thinks of "culture" as a means of shifting the ongoing class struggle from the streets onto the individual mind. In the theater of consciousness, he imagines high culture subordinating the materials of lay culture to an exalted conception of Englishness. Rather than equating "anarchy" with the "carnivalesque," as White's data suggests respectable people did, Arnold aestheticizes the terms of domination and subordination. Though he calls them "Hebraism" and "Hellenism," the model exhibits the familiar behavior of dark underworld forces that both disrupt the reigning (rational) order of things and prompt that order to achieve greater clarity and comprehensiveness.

There are two points to be made from these discussions of the relationship between violence and modernity. First, it is clear that a discourse is in place ensuring that any form of resistance can be regarded as something requiring expression, form – acculturation. Second, such a hegemonic culture constantly repositions "other" cultures and viewpoints in a negative relationship to rational middle-class masculinity. This model of culture ensures the relation of dominance and subordination. At the same time, it also makes the dominant culture terribly dependent on its Others by locking the two into a mutually-defining relationship. In such a relationship, finally, even the dominant culture cannot stay the same; it is always negotiating, renegotiating, and absorbing forms of difference.

Contemporary culture: the art of politics

To move from Matthew Arnold to "Vietnam" may seem to require an enormous conceptual leap, but in actuality we must pause deliberately in order to pay due respect to the passage of nearly a century before rushing into what has – rightly or wrongly – come to be called the postmodern condition. Indeed, John Carlos Rowe demonstrates how the cultural uses of the Vietnam War offer a curious case of Arnoldian wish-fulfillment. Americans today – not just those with specialized knowledge of ideology, but disgruntled citizens of all sorts – have become sensitive to the political information they consume with education, advertising, and many of the activities that have to do with making a living. Of the ideology governing this dimension of their lives, or even prime-time television programs, great numbers of people can become intensely critical and still carry on the work demanded of members of our society. But how many of these same people still bring anywhere near the same degree of critical consciousness to their leisuretime activities? Rowe's essay suggests that ideology is doing its work – perhaps its most powerful work – through the information absorbed or the thing we do when we think we are free.

This possibility assumes new strategic importance under conditions of late capitalism. What is potentially at stake in the fuss over postmodernism is, we believe, the larger question raised by post-structuralist theory and encapsulated in Baudrillard's figure of the simulacrum: "The very definition of the real has become: that of which it is possible to give an equivalent reproduction. . . . The real is not only what can be reproduced, but that which is always already reproduced" (1983: 146). The postmodern condition (if such it can accurately be called) clearly announces the loss of a traditional basis for referentiality, which of course portends the end of the distinction between base and superstructure as anything more or less than discourse. This possibility has thoroughly unsettled theories of culture which hold that ideology derives from and refers back to essentialized categories of political economy (such as "labor" and "class"). Understood in terms of a Gramscian concept of hegemony, however, ideology acquires new importance as a result of the shift of productive labor to Asia and elsewhere, and the emergence of a massive class of people who produce, manage, and distribute information. Rowe's study of the revision of the Vietnam experience two decades after the "fact" is executed with a sense that our use of the analytical techniques acquired from a literary training is more important than ever before. If one accepts the assumptions of Baudrillard's similacrum, then we are no longer once removed from political relations in working with printed matter or film. We are bringing interpretative skills directly to bear on it. Culture and power are one.

If Rowe's thesis derives from Arnold's model of culture in this general historical sense, it is connected in a more specific way as well. Rowe tells a tale of historiography, or how the Vietnam War was remade in various films – documentaries as well as fictionalized accounts of the conflict – in which the qualities of anti-war protesters (and even the Vietcong) were incorporated into characters authorizing more recent domestic and foreign policy. The centerpiece of the essay is Rowe's discussion of *Rambo*:

> By associating Rambo with the frontier-spirit, Transcendentalist self-reliance, twentieth-century neoruralism, the anti-war and civil rights movements, the revolutionary zeal of the National Liberation Front, and the politicized Vietnam veteran, Stallone's two films incorporate these different and clearly critical discourses into the general revisionary view of the Vietnam War. . . . Aligning himself with all of those, Rambo directs his rage against the military–industrial complex, thereby diverting genuine dissent into an affirmation of the very nationalism that took us to Vietnam in the first place.

The idea that this is just what it seems, unreflective nationalism, is belied, in Rowe's account, by the very scenes that declare it so: Stallone smashing data-gathering machines, or declaring by the mute violence of his hero that actions speak louder than words. But in fact his embodiment of certain features of the Vietnam conflict comes at the end of a process that has transformed the conflict substantially and fine-tuned it to serve specific ideological ends. If Rambo embodies both veteran and protester in an authentic American who goes up against information-oriented technology, then years ago such films as *Deliverance* and *Southern Comfort* transformed the foreign devil into an enemy within our borders. This enemy haunted our rural bywaters and national boundaries to stalk the true American, much as his Vietnamese counterpart stalked American foot-soldiers in Asia.

Such Orientalism (as Rowe, following Edward Said, identifies it) has its roots in the earlier form of cultural imperialism for which Arnold was one of the most articulate spokesmen. The power of Rowe's argument is to show why we can no longer blithely distinguish political action from the recombination of coded features in and through the conflicts portrayed on news programs and the popular screen. Not if the screen itself has become the arena of conflict. As Rowe more eloquently puts it.

> The violence of representation operates in the rhetorical uncanniness of American ideology, which can and does turn just about anything to its

> purposes with a sensitivity for the connotative and figural qualities of language unmatched by the subtlest artist, the cleverest critic. American ideology in the aftermath of the Vietnam War has managed to live up to Jean-Francois Lyotard's definition of the postmodern: ''the presentation of the unpresentable.'' . . . ''Vietnam'' as unpresentable? In a sense, that is, of course, the case for America in the 1980s; ''Vietnam'' remains the name that cannot be properly uttered, even though we continue to translate it into very presentable terms, very recognizable myths and discourses. What is finally *unpresentable* is the immorality of our conduct in Vietnam. . . .

Lucia Folena's study of Reaganite rhetoric carries this critique to contemporary American policy in Central America: ''Naming is renaming, defining (marked linguistic boundaries), translating Sandinistas into Stalinistas and *contras* into freedom fighters, amounts to waging the first offensive in the war of cultural representation.''

Folena takes us back to Jacobean England and ''the first great Scottish witch-hunt'' to demonstrate how this particular kind of violence serves the interests of an imperialist nation-state. The point of King James's demonizing of witches was not to falsify the nature of the Other but rather to situate the king as the subject of discourse that opposed the Devil. The objective was an entirely discursive one: ''the centralization of the monarch trope and the consequent displacement and marginalization of the Kirk.'' By claiming the Devil as his true enemy, Folena suggests, James not only identified himself as the true enemy of the Devil but also avoided direct conflict with members of the nobility who were contesting his power.

In mounting an argument that identifies Reaganite strategies of representing Nicaragua with James's practice of demonizing the Other, however, Folena appears to take issue with the idea that rhetoric became the major instrument of political power only recently. Indeed, her thesis echoes Jed's, contending that the difference between literary and political language, like the difference between any use of language and events that occur elsewhere, is one that is discursively produced. But this does not mean, as some warn, the loss of ''literature'' and (what is probably more to the point) the devaluation of our training and hermeneutical function. To the contrary, Folena suggests, ''if King James's *Daemonologie* is literature . . . so are Ronald Reagan's speeches, and they should be read as such.'' For were we to abandon attempts, however sophisticated, to distinguish such modes of writing, the philological strategies at work in political utterances would soon become apparent. To demonstrate her point, Folena proceeds through the layers of Reagan's performance as official spokesman for Central American policy, revealing the tautologies and paradoxes which go into making a simple human truth – in this case, the righteousness of American imperialism. Like the performance, the purpose served by demonizing the Other is no simpler or more direct than that of James I:

> Foreign policy is a figure of domestic politics. For there is an internal Other. There are those who "are confused" and "refuse to understand"; they are "the new isolationists," and "would yield to the temptation to do nothing" to restrain "the aggressor's appetite." The real enemy in this sense is the Congress: "the opponents [in a vote on *contra* aid] in the Congress of ours, who have opposed our trying to continue helping those people, they really are voting to have a totalitarian Marxist–Leninist government in America, and there's no way for them to disguise it."

The violence done to the Other, in Folena's analysis, ultimately finds the objective elsewhere, and the people who actually suffer the violence do so in the name of a conflict that is much closer to home.

Violence in theory as violence in fact

If these essays can be said to demonstrate a single point, it is this: that a class of people cannot produce themselves as a ruling class without setting themselves off against certain Others. Their hegemony entails possession of the key cultural terms determining what are the right and wrong ways to be a human being. With this in mind, we have tried to provide some sense of the detailed process by which certain people, a relatively small group, at different times produced the Other in specific ways. In so articulating our project as a collective project, we want to insist that what we have offered is a story about the production of a culture-specific subject and only a very partial one at that; it suggests very few of what we believe were the myriad ways in which differences were suppressed in this process and positioned in a negative relationship to the ruling-class self. In this respect, our narrative will inevitably reproduce the very behavior it set out to historicize. It will exclude points of view that are not of the dominant race, gender, class, and ethnic group. This is why we have chosen to conclude our collection with an essay by Teresa de Lauretis.

She brings our subject of history into the present day (as indeed do Rowe and Folena) and links the mode of domination to the practices of those of us who write about written culture. She is particularly concerned to make us aware of the way in which theories of representation themselves maintain the hegemony of a specific subject. Thus the words we use to represent the subjects and objects of violence are part and parcel of events themselves. Naming the perpetrator and victim is implicated at every point in the event – what we feel it is and how it makes us feel. Though the impulse is to posit a one-to-one relation of discursive to social events, de Lauretis explains,

> once that relation is instated, once a connection is assumed between violence and rhetoric, the connection will appear to be reversible. From the Foucauldian notion of a rhetoric of violence, an order of language which speaks violence – names certain behaviors and events as violent, but

not others, and constructs objects and subjects of violence, and hence violence as a social fact – it is easy to slide into the reverse notion of a language which, itself, produces violence. But if violence is in language, before if not regardless of its concrete occurrences in the world, then there is also a violence of rhetoric, or what Derrida has called "the violence of the letter."

De Lauretis cites a study of "family violence" to show the unwillingness of medical personnel to link the terms or, in linking them, to intervene in a situation where women's and children's lives and limbs are in jeopardy. Rather than think of the family as something violent, the study concludes, medical professionals prefer to take women who were hit and think of them as "battered," thus coercing them back into a destructive situation. It is not the family but the woman who is deemed out of order. In analyzing the reversibility of violence, the authors of the study differ from the medical men whose linguistic behavior they criticize. They refuse to see violence as the breakdown of social order, identifying it instead with the maintenance of a pattern of dominance. This idea of violence as representation is not an easy one for most academics to accept. It implies that whenever we speak for someone else we are inscribing her with our own (implicitly masculine) idea of order. Indeed, as de Lauretis shows, the most self-conscious theorists of culture, from Lévi-Strauss and Girard to Foucault and Derrida, have used gender in precisely this way. As the recent critique of academic feminism has revealed, furthermore, in presuming to speak for "woman," feminist theory sometimes resembles the very thing it hates and suppresses differences of class, age, and ethnicity, among others.

But to renounce theory, as some have attempted to do, is outrageously cynical, especially if what de Lauretis says is true. Backing away from the power of representation simply allows one to continue to represent others as if one's view is either "fact" or else simply one's opinion. Precisely where we think we are relying on the most obvious existential fact, on the one hand, or on our most deeply held feelings, on the other, is where we are being most ethnocentric, our thinking the most passive vehicle of ideology. Assumptions instilled during the early phases of our socialization, those we require to make the world intelligible from our specific position within that world, speak through us then – and maintain an order of things that we hold to be unquestionably real. By virtue of being academics, we are therefore directly involved in the violence of representation. As we hope the essays in this volume demonstrate, there is no way to position ourselves outside of an ongoing struggle among viewpoints. Neither the deepest recesses of the female psyche nor the loftiest pinnacles of art afford us any refuge; to position oneself outside of the struggle in the manner of Arnold or Freud is simply to carry the struggle elsewhere – into new territories where it is more difficult to see our scholarly and pedagogical objectives for the political objectives they are.

Condemned to power, then, it seems only reasonable to acknowledge who we are and what we do. This requires us to recognize the authority of Jane Eyre and the kinship between the violence of her discourse and our own. After all, we who work in certain areas of the humanities and social sciences are in the business of naming the Self and its Others. Whether we do this work self-consciously or not, there is no way around it. That is what we do when we read texts, roles, performances, models, minds, and deeds of all sorts, theoretical and concrete, fictive and empirical, past and present. And like the medical personnel in de Lauretis's study, we are, in naming, also constituting information as an event, transaction, or relationship. We are positioning it within our culture and in relation to ourselves. As a group, we thus play a central role in the ongoing history in which identities are produced, contested, negotiated, renegotiated, and ultimately revised. We exercise the very form of power which, as we have argued, gained centrality as part of modern imperialism. Thus the question is not how to renounce that power. There is no position of non-power from which we can write and teach. The question is rather how to become politically self-conscious, which requires political histories of our position and of the kind of work we do.

References

Baudrillard, Jean (1983) *Simulations*. Trans. Paul Foss, Paul Patton, and Phillip Beitchman. New York: Semiotext(e).

Brontë, Charlotte (1975) *Jane Eyre*. Ed. Margaret Smith. Oxford: Oxford University Press.

Foucault, Michel (1978) *The History of Sexuality*. Vol. I, *An Introduction*. Trans. Robert Hurley. New York: Pantheon.

Part One

Early Modern Culture: Putting the Politics Back into Poetics

1

The scene of tyranny
Violence and the humanistic tradition

Stephanie Jed

Narratives of political violence occupy a notable space in the writings of fifteenth- and sixteenth-century humanists. In particular, narratives representing the means by which tyranny is either maintained or overthrown were of interest to the Florentine humanists and exercised a certain amount of authority over the relations to politics and writing they imagined for themselves. But the question of whether or not represented violence had any effect on the interpretative and methodological procedures of those humanistic readers and writers who found such representations appealing is seldom raised.[1] Is it possible to identify an enactment of these figures of violence and power in the interpretative activities of the humanists?

Contemporary philosophic thought has presented the urgent need to investigate the humanistic tradition for signs of an unacknowledged violence.[2] This need arises from the fact that, while the humanistic tradition is often proposed as an exemplary site of interpretation, little attention is paid to the conditions under which this tradition is reproduced and transmitted. One might well suspect, then, that an exercise of domination and suppression may be implicated in the very conditions of humanistic codification. This essay examines two specific narrative representations of violence, coded by humanists for ''literary'' transmission. The first is the account of one tyrant who covertly advises another to maintain his tyranny by ordering the mass decapitation of noble citizens; the second is the narrative of a young fifteenth-century Latin scholar who, induced by his reading of Latin literature, carries out a political assassination. In each case, the narrative in which violence is figured is coded in such as way as to disguise the violence it exercises, as a narrative, upon the scene of writing.

A message of violence and the humanistic code

In the fifth book of Herodotus' *Histories*, Periander, the tyrant of Corinth, sends a messenger to Thrasybulus, the tyrant of Miletus, to find out how he might best govern his state (V.92). In response to this query, Thrasybulus leads the messenger to a field and cuts off the tallest standing ears of grain, while questioning the messenger about his arrival from Corinth. The messenger then returns to Corinth and reports to Periander how he received no response from Thrasybulus and how Thrasybulus seemed crazy to him for the way in which he had destroyed his grain. Upon hearing the messenger's report, Periander knows he has been advised to murder his most powerful subjects: the cutting of the tallest standing grain in Miletus is understood in Corinth as a sign to decapitate the most prominent citizens.

This tale of tyranny is transmitted and codified by a tradition which finds metaphorical stories about political violence central to its own concerns. Aristotle, for example, reminds his readers of the tale in his discussion of ostracism as a means of maintaining equality in the state (III.13). Livy retells it to fit the needs of early Roman history, changing the names of the cities to Gabii and Rome, transforming Thrasybulus and Periander into the ''Italian'' tyrants, Tarquinius Superbus and Sextus Tarquinius, and substituting the tallest ''heads'' of poppies for the tallest standing ears of grain (I.54). Shakespeare has a gardener deliver advice about the tallest standing plants:

> Go thou and, like an executioner,
> Cut off the heads of too fast growing sprays
> That look too lofty in our commonwealth.
> All must be even in our government. (3.4)

And in his commentary on Tacitus, Filippo Cavriani writes that the best strategy for the suppression of rebellions is the decapitation of poppies in the garden:

> Tiensi, che la vera strada d'opprimere le ribellioni sia quella mostrataci primieramente da Aristobolo: et imitata poi da Tarquinio di Sesto Tarquinio padre, col gettare a terra i capi dei piu belli, et de i piu grossi papaveri dell'horto (522)
> (It is held that the best way to suppress rebellions is the one demonstrated first by Thrasybulus[3] and then imitated by Tarquin, the father of Sextus Tarquinius, who struck to the ground the heads of the tallest and most beautiful poppies in the garden)

Although the transmission of the narrative of the tyrant in the field is actualized by distinct agents operating in distinct historical moments, the accumulation of these instances produces an overall rhetorical effect. The entire series of tyrants in the field is transmitted to the scholar of western tradition in one package, as if it were one text punctuated by the recurrence of this tyrannical motif.[4] What begins in Herodotus as the establishment of an analogic relation between the tallest standing grain and the most prominent citizens is eventually transformed, by virtue of the acquired familiarity of the story, into a *topos* of humanistic discourse about tyrants.

The creation of this *topos*, of course, does not occur in a social and political vacuum, but results from the narrative's codification by a particular class of readers in whose interests it is to codify the tyrant's "decapitation" of ears of grain as a referent of its own conceptualizations of power and violence (White, 1982a: 288). Thus, every allusion to the tyrant's message of violence refers not only to Herodotus' text, but also to the history of reading and writing practices by means of which Herodotus' text is transmitted and codified. Humanism, the ancestor of our modern-day humanities,[5] discourages us from scrutinizing the specificity of humanistic *topoi* to a particular kind of literacy and cultural code; any attempt, therefore, to locate Thrasybulus' message of violence in the cultural code of humanists who transmit this narrative seems, at first, illogical and absurd. And yet, the *topos* of the tyrant in the field, inasmuch as it thematizes, in its representation, the relation between Thrasybulus' message and the code of tyrants, invites precisely this type of scrutiny. For the mode of communication between tyrants is, in some respects, analogous to the mode of communication between humanists.

Seen from a semiotic perspective, the tyrant's "decapitation" of ears of grain or poppies forms part of a complex communicative process: a coded message is transmitted from one tyrant to another and the narrative of this exchange is transmitted from one reader to another. Furthermore, just as Thrasybulus knows that the coded message transmitted by Periander is to be deciphered with reference to the maintenance of tyranny, so the humanist knows that the codified texts transmitted by other humanists are to be interpreted so as to perpetuate a "republic of letters" regulated by a particular class of readers and writers.[6] But there is more than just a simple analogy at work here; if Thrasybulus' message of violence is implicated in the code of tyrants, then it becomes important to investigate the relation of these coding processes to one another. Evidence about this relation emerges, if we consider the reproduction of this narrative as an active response to Thrasybulus' message of violence.

The tyrants' coding process is entirely figured within the confines of the narrative. The tyrant (whether Thrasybulus or Tarquinius Superbus) communicates his advice to decapitate the most prominent citizens by means of a metaphorical violence committed against the grain (or poppies). And the violence takes place at the level of representation: the decapitation of citizens remains within the limits of verbal configuration. Once this narrative is transmitted, however, there is no insurance against the eruption of the tyrants' violence outside the limits of textual representation. For, at this point, the violence of the tyrants is embedded within a cultural practice by means of which the narrative is transmitted and reproduced.

In the transmission of this text from humanist to humanist, the sender and the addressee are human interpretants who, stimulated by their study of the classics, are moved to enact these texts *outside*, as a part of their daily activity.[7] Just as Thrasybulus transmits a message of metaphorical decapitation, so that

Periander will enact the metaphor on the heads of citizens, so humanists, in their codification of classical learning, transmit figural messages to other humanists in the hope that they will enact the figures in their own lives. For the most part, this enactment takes the form of discussions, editions, reproductions, interpretations, and citations of the classical texts. But in cases of narratives of "tyranny" and "freedom," classical texts have been enacted in real political propaganda, real political opposition, and even in real violence, real murders.

Humanists are incited by their reading of classical literature to commit real acts of violence. And this violence may seem, at first, to be incompatible with the tranquility of the scriptorium, the site of the humanist's engagement in scholarly, noble, non-violent activity. But what about the narratives of rape, murder, and violent contests between forces of "tyranny" and "freedom" which are central to the corpus of texts codified by the humanistic tradition? The making of the humanistic code cannot be divorced from the recurrence of such figures as the decapitator of grain, the vindicator of violated honor, the scholarly tyrant-slayer, etc. And if tyrants are able to maintain their "tyranny" by transmitting coded messages of violence, some vestige of this violence must remain in the humanists' codification of this tale. As heirs to this tradition, we, modernday humanists, might want to examine the violence in our own culture as a possible by-product of our "literary" activities. For the modern "literary" discipline continues to enact (in public lectures, conferences, interpretative essays, etc.) figures codified by the humanists without acknowledging the violence of the code.

The rupture between "literature" and "politics"

In 1456, Cosimo de' Medici commissioned Vespasiano da Bisticci, the merchant "prince of booksellers," to supply him with 200 books for a new library at the Abbey of Fiesole.[8] Forty-five scribes, experienced in copying a repertoire of texts that had been in demand for over a hundred years, filled the order in twenty-two months. In this rapid and artificial production of a library (representing the synchronic efforts of a book-making industry rather than the interests of successive generations of readers), a precedent was set for the formation of the library as a symbolic instrument of the power of the state (Petrucci 1983: 547–51). Subsequently, Vespasiano supplied other rulers, including Mattia Corvino, King of Hungary, and Federico da Montefeltro, Duke of Urbino, with "humanistic libraries" as symbols of their power. Although competition from the printing press eventually forced him out of business in 1480, Vespasiano's entrepreneurial activities had a lasting cultural legacy.

The creation of the "humanistic library," coupled with the avant-garde interests of humanistic scholars in the linguistic and formal properties of "authentic" Roman Latin and script, helped to formalize an already operative cultural practice: a certain kind of writing, coded as belonging to a humanistic library, was isolated from the kinds of writing practiced in mercantile and diplomatic offices, in courts of law, and in shops, by tax collectors, notaries, etc. By means of this isolation, not only was humanistic "literature" detached from the activity or process of its production, but signs of the relation between "literature" and other kinds of lettering of society were tacitly suppressed. As a consequence, the consideration of humanism as a particular practice of literacy to be seen in relation to others came to be overshadowed by the humanistic ambition to dominate the scene of writing with exclusively "literary" models. This suppression and domination implicit in humanistic cultural practices can be effectively illustrated by a humanistic representation of tyrannicide in which the only kind of writing represented by the narrator is the Latin writing inducing three young scholars to commit murder.

Cola Montano, an ambitious scholar ("uomo litterato e ambizioso"),[9] taught Latin in Milan to the noble youth of the city. He became particularly intimate with three young students, G. Lampognano, C. Visconti, and G. Olgiati, and he made them swear that, as soon as they were old enough, they would liberate their city from the tyranny of the duke, Galeazzo Maria Sforza. Galeazzo was lascivious and cruel. Not only did he rape noble women (including women associated with Visconti and Olgiati); he received even more pleasure in publicizing their dishonor. He wasn't satisfied to kill men, unless he had killed them by some cruel means. He lived with the reputation of having killed his own mother.

The three young students made plans to kill the duke. They hoped that once the duke was slain, the patricians, followed by the people, would take up arms against the duchess and princes of the dukedom. Their hopes were founded on the hunger of the people and on their plan to turn some property of the princes over to the people as their booty in the struggle for liberty.

On December 26, 1476, in the church of Santo Stefano, the three young Latinists slew the duke. Two of the conspirators, Lampognano and Visconti, were also killed. And Olgiati, after hiding out for a couple of days hoping for an insurrection, was caught, confessed, and was sentenced to be tortured and quartered alive. His bravery was such that even as he faced his executioner, he delivered a long oration in Latin, because he was "litterato." The speech ended with his aspiration to achieve fame for having slain a tyrant. His death would be bitter, but his fame would be perpetual, because he would join the ranks of all of the other tyrant-slayers of Greek and Roman literature.

To narrate the event of the assassination of Galeazzo Maria like this, as Machiavelli does, is to interpret the event according to the criteria of the humanistic code. The intelligibility of the narration depends upon the degree to which it makes reference to the narratives of tyrannicide transmitted from classical antiquity. Thus, Galeazzo Maria is figured as the classical tyrant who rapes noble women, devises cruel tortures, and kills his own mother. And Olgiati is figured as a classical tyrant-slayer, whose yearning for freedom and willingness to sacrifice his own life to that end originates in the study of classical literature. This narration is written for readers who will recognize the typical features of the tyrant and tyrant-slayer because of their own study of classical literature; it is a narrative written from the perspective of Girolamo Olgiati, the Latin scholar who aspired to have his name and deed transmitted to future generations of literary scholars.

Even if his attempt to win freedom for Milan was a failure, Olgiati was successful in recording his name for posterity; scholars are still reading and talking about him. The transmission and codification of Olgiati's story serve to perpetuate the humanistic practice of privileging classical learning as the true source of political understanding and aspirations for freedom. Even though Olgiati's classical learning was clearly insufficient to the carrying out of the goal of liberation, his "literary" code continues to be transmitted. Passed on from humanist to humanist, there is no need to mark the humanistic code as one code among many struggling to dominate the scene of writing in fifteenth-century Florence (or in twentieth-century USA). In Olgiati's ambition to make his code prevail, we may catch a glimpse of the politics of humanism, which, having lost its power to represent "politics," still keeps a tight grip on our own "literary" enterprise.

One feature of this politics of humanism can be detected in the difficulty modern humanists have in explaining acts of political violence, committed by fifteenth- and sixteenth-century scholars who were merely following the prescription of a "literary" code. To explain the failure of this humanistic violence to achieve its political goal of liberation, modern evaluations generally posit a rupture between "literature" and "politics." According to this view, humanistic projects of violence fail in their assessment of political "reality," because too much "literary" activity removes these humanists from contact with real political concerns and prevents them from achieving an adequate grasp on political facts. In his treatment of Lorenzino de' Medici's assassination of the Duke Alessandro, for example, the political historian Rudolf von Albertini claims that Lorenzino's erudition caused him to substitute a "literary model" for "political reality." And extending his critique to an entire generation of humanistic opponents to Medicean rule, Albertini identifies the rupture between "literature" and "politics" in the consciousness of humanists as the operative cause of their political self-deception:

> The years of exile, that most of them had spent in literary work and in translations from the ancient authors, had made a full evaluation of the Italian and Florentine political situation impossible. Memories and hopes, little by little removed from reality, had assumed a literary hue. (1955: 211)

But to indicate a rupture between ''literature'' and ''politics'' is to operate within the humanistic tradition of isolating ''literature'' from other kinds of writing, in and of itself a political move; once ''literature'' is isolated from all that is ''political,'' it becomes difficult to consider writing emanated from political institutions under the rubric of ''literature.'' By the same token, once ''politics'' has been defined as a ''non-literary'' phenomenon, it becomes difficult to consider the ''politics'' involved in determining which kinds of writing may qualify as ''literature'' and which may not.[10] The literature of politics and the politics of literature are deftly suppressed by Albertini's identification of the isolation of humanistic activities from political ''reality.'' Albertini's move is descended from the humanists' own attempts to organize political ''reality'' according to a ''literary'' code.

The politics of humanism

In his pioneering essay ''Rhetoric and Politics in Italian Humanism,'' published in 1937 in the first volume of the *Journal of the Warburg Institute*, Delio Cantimori underlines the importance of considering the politics of humanistic culture. In contrast to Albertini, who represents the ''memories and hopes'' of the humanists as irrevocably severed from the ''Italian and Florentine political situation,'' Cantimori perceives a continuity or nexus between humanistic ''nostalgia'' for the Roman republic and the ''realistic politics'' of sixteenth-century Florence (1937b: 84–5). Citing, in particular, material from Antonio Brucioli's *Dialogi della morale filosofia*, Cantimori concludes that

> there is no schism in Italian humanism between ''rhetoric'' and ''literature'' on the one hand and ''politics'' on the other, but . . . politics draw their sap from literature which in turn they fertilize. (ibid.: 102)

In the case of Olgiati, however, Cantimori's critique of Olgiati's tyrannicide essentially coincides with the charges leveled by Albertini against the anti-Medicean humanists of the 1530s; in Cantimori's view as well, Olgiati's political aspirations are so dominated by ''literary'' models as to be inadequate to the confrontation with a ''real'' situation. None the less, according to Cantimori, a ''political reality'' is never ''missing'' from the consciousness of a historical protagonist.

In his essay ''Il caso di Boscoli e la vita del Rinascimento,'' Cantimori cites a variant in the narrative of inadequate political planning; he sets up a contrast between two humanist/tyrant-slayers – Olgiati and Pier Paolo Boscoli – and judges each according to the degree of his adherence to the humanistic code. Olgiati, he claims, remains until his last breath entirely faithful to the humanistic credo that the codified texts of classical antiquity provide the only appropriate mediation between individuals and their socio-political situations. Thus, Olgiati's perceptions and ideals remain within the limits of a "literary construct" bearing

> no real connection to the concrete historical development of politics of the time. Desire for political freedom, love of the fatherland, of the city, hatred of the tyrant, enthusiasm for Brutus, Sallust, study of the philosophers – all of it bookish and not ripened in life – is enough for Olgiati. (Cantimori 1927: 249)

His enthusiasm for liberty lives freely within the confines of the humanistic code, because it is an aesthetic enthusiasm. But once such enthusiasm collides with the "more profound world" – that is, "concrete reality" – it loses its vitality and dies (1927: 251).

In Boscoli's enthusiasm, by contrast, there is a glimmer of consciousness that the humanistic code is not enough to inform concrete projects of political change. After his attempt in 1512 against the lives of Giulio, Giovanni, and Giuliano de' Medici, awaiting execution, Boscoli finds his aesthetic-literary vision of politics to be of little comfort. Boscoli's last words show him feeling the inadequacy of his classical studies to the challenge of political action. In the minutes before his execution, unlike Olgiati who speaks in Latin about his "literary" mentors, Boscoli rejects the myth of Brutus' heroism. He rejects as mere rhetoric the classical illusions which had deceptively occupied his mind to the exclusions of other mediations of reality, and he cries to his confessor: "Take Brutus away from my head, so that I can take this last step entirely as a Christian" (1927: 250; della Robbia 1842: 290). Boscoli's conversion from Brutus to Christ is a superficial one, according to Cantimori, unsustained by a critique of either humanism or Christianity. But his awareness of the inadequacy of the humanistic code to withstand a collision with the "more profound world" prepares the way for the heresies of Pico, Savonarola, and Bruno who, in the staging of this collision, became "founders of the modern world" (Cantimori 1927: 255).

Cantimori's representation of the figures of Boscoli and Olgiati and the relation of their consciousness to the humanistic code reflects the contorted situation of Italy in 1927 and Cantimori's own attempt to constitute, in his own political activities, an existential continuity between the scholarly study of humanistic violence and the problems of political consciousness particular to fascist Italy.[11] Thus, the figures of Olgiati, Boscoli, and the protagonists of the Reformation are arranged, in a graded succession, according to their ability to make their own philosophic, political, and religious codes survive a confrontation with "reality." Those who are able to leave the mark of their aesthetic enthusiasm on the world outside the "republic of letters" can be called real agents of political change.

According to this view, Olgiati is the least effective, in that he, at no time, gains consciousness that the classical narratives are not powerful enough to leave their impression on "reality." And Boscoli, in this graded list, may be seen as a precursor of the Reformation heroes, inasmuch as his enactment of the classical narratives of tyrannicide showed him how inadequate the humanistic code is in a confrontation with "reality." The ability to incorporate contrasting codes into projects of political change is clearly a value for Cantimori, whose research on humanism and heretical movements informed his own political activity. It is difficult, for example, to separate the development of Cantimori's thought and scholarship on humanism in the 1920s and 1930s from the essays he wrote, in this period, on political propaganda or from the ideological orientation of the journals to which he contributed; through a comparison and confrontation of the codes of humanism and fascism, Cantimori was able to sustain the integration of "bookish" research with his own life of political-cultural militancy.[12]

From literacy to literature

Underlying Cantimori's evaluation of the politics of humanistic rhetoric and the relation of humanistic *topoi* to the "realistic decisions derived from them" (1937b: 91) is an implicit acknowledgement that even when "literature" alone is inadequate to the understanding of political "reality," political "reality" is made up of various "literatures." Just as the political "reality" of Italy in 1937 was, at least in part, constituted by the "literature" emanating from fascist institutions, so the political "reality" of fifteenth- and sixteenth-century Italy was constituted by the kinds of "literature" emanating from political and economic institutions of that period in which writing was extended to new social groups and new functions. To take Cantimori's intuition even further, political "reality" is nothing but the clashing and mixing of different "literary" codes materialized in the production and transmission of verbal messages. That the humanistic code has come today to exercise an almost exclusive right to call itself "literature" is not the least of the achievements of humanistic politics: by concealing the social specificity of the humanistic code to a particular kind of "literacy" or cultural practice, humanism obstructs the access of other written codes to representation of the sign "literature."[13] For example, neither the "literature" of fascist Italy nor fifteenth- and sixteenth-century archival documents have yet, for the most part, made their way (except as information digested by historians) into our courses on that category of writing – "literature" – which claims to transcend our everyday concerns. And we find it difficult to connect the narrative of the violence of tyrants codified by humanism as "literature" to the violence represented daily in court records, in documents emanating from the Pentagon, or in a communique issued by the death squads in El Salvador. Yet, in each case, we are dealing with pieces of literature, though generated by different codes.

Machiavelli's account of Olgiati's execution offers a crucial moment of ambiguity in which we may view the transformation of humanistic literacy into what we call today "literature:"

> Girolamo was 23 years old, and he was no less courageous in dying than he had been in performing [the assassination]; finding himself naked and his executioner before him with the knife in hand to wound him, he said these words in the Latin language, because he was *litterato*: "Mors acerba, fama perpetua, stabit vetus memoria facti" (1962: 507; my italics)

The key term here is *litterato*, a sign we associate today with literary elites. But the term *litterato* also refers to a particular brand of literacy – to the ability to read and write in Latin. This brand of literacy is produced by the transmission and reproduction of classical literature to and for a particular social class which makes this literature the focus of its identity and social relations. But Olgiati's literacy is not the only brand in circulation in 1476 (or 1512, 1537, 1985, etc.) The production of writing is not the exclusive practice of Latin scholars. Other types of literacy – such as the ability to write Italian for commercial, diplomatic, and generally non-classical purposes – would not, however, mark a person as *litterato*. Even Leonardo da Vinci, a great hero of humanistic literacy today, knew that many men looked down on him as an *omo sanza lettere*, because he knew no Latin (Garin 1975: 59, 79, 99). From a perspective which considers various kinds of literacy, "reality" would not be understood as a world without writing, but rather a world written or lettered by different codes – not by the institutions produced by the transmission of classical literature but by the political and economic institutions whose "lettering" of society constitutes a different kind of regulation.[14] What, then, prevents these other brands of literacy from becoming the objects of "literary" study?

Conclusion: violence and the "work of resemblance" (Ricoeur 1977)

Although we are more accustomed to associating humanism with scholarly activity than with violence, there are structural similarities between the code of tyrants and the code of humanists which unmask the violence of humanistic cultural practices. The intelligibility of the tyrant's message, for example, depends upon its reference to a code shared by tyrants, just as the transmission of the narrative depends upon its reference to the code of humanism, shared by a particular class of fifteenth- and sixteenth-century readers whose cultural activity revolves around the study and reproduction of Greek and Latin literature (Eco 1979: 78–9). In like manner, the exclusion of the messenger from the construction of the tyrants' code is replicated in the exclusion from humanistic elites of all those who are ignorant of Greek and Latin. And if it is the exclusive power of the tyrants to understand the meaning of Thrasybulus' cutting gesture, it is also the exclusive power of humanists to reproduce Thrasybulus' gesture as a *topos* of tyrannical violence.

Both the tale of Thrasybulus and its codification by humanists are products of the "work of resemblance." The secret communication between tyrants could not have been completed if the code of tyrants had not worked to establish a resemblance between ears of grain and citizens. In the same way, this tale would not have become a part of the humanistic canon if the code of humanists had not worked to make the code of tyrants "congenial" to its own conceptualizations of politics and violence (White 1982a: 289). To disclose this continuity or resemblance between the figural content of Thrasybulus' message and the cultural operation by means of which the figure is transmitted is to draw attention to the means by which the *topos* of the tyrant in the field becomes a trope or prescription for the enactment of violence. Although it is perhaps more intriguing to consider how this "literary" violence functions as an incitement to political violence *outside* the text, the urgent task at hand is to discover how this violence is sublimated *inside* today's practice of literary interpretation, inside the "republic of letters."[15]

There seems to be a tacit prohibition in humanistic studies, corresponding to the incest taboo: never identify the metaphorical relation or relation of similarity between the literary figure and the interpretative code of which it comes to form a part. Or, in terms of our examples here: never identify how the violence of the tyrant in the field, codified and transmitted by the humanistic tradition, may have influenced the workings of the code. Linguistic theory has taken the lead in making this prohibited identification, by disclosing the structural similarity between the "expression plane" and the "content plane" of linguistic messages. Following the scheme of Hjelmslev (and Jameson's literary application of it), we can see, for example, how both the humanistic code and the codified figure of the humanist/tyrant-slayer are "structured in quite analogous fashions" (Hjelmslev 1963: 60; Jameson 1981: 145–50). And in his classic essay, "Two Aspects of Language and Two Types of Aphasic Disturbances," Jakobson extended this structural analogy, noting that the literary researcher tends to focus on metaphor, because a metalanguage is more easily constituted in a relation of similarity to the object of interpretation:

> when constituting a metalanguage to interpret tropes, the researcher possesses more homogeneous means to handle metaphor, whereas metonymy, based on a different principle, easily defies interpretation. (1956: 81)

In the field of tropology, Hayden White has shown that a certain amount of suppression may be detected in the assimilation of linguistic signs to the metalinguistic structures of interpretative work; by virtue of their similarity to the language of the interpreter,

> certain sign-systems are privileged as necessary, even natural ways of recognizing a 'meaning' in things and others are suppressed, ignored, or hidden in the process of representation. (1982a: 288)

And in the case of the humanistic tradition, the same dynamics of assimilation is at work and, of course, the same consequence of suppression and violence. Part of the dream of tyranny is the desire to suppress the transmission of messages which refer to other codes besides the code of tyrants. If the story of the code of tyrants is, in turn, codified by humanists, it is because humanists share this dream and ambition to make the humanistic code prevail upon political "reality."

In what did Montano's ambition consist and how may we relate it to his brand of literacy – characterized by the study of Greek and Latin literature? Cola Montano's ambition was to transmit the tools of humanistic literacy to the noble youth of the city in the hopes that their conceptualizations of politics and violence would be grounded in reference to the humanistic code. In semiotic terms Montano was successful in that three of his students, Olgiati, Lampognano, and Visconti, projected the form of the classical tyrant on the person of the duke and worked to assimilate the political codes operative in Milan in 1476 to the liberation narrative of Brutus, codified by the humanistic tradition. Although this "work of resemblance" failed inasmuch as the liberation of Milan did not ensue upon the slaying of the duke, Cola Montano's ambition was nonetheless realized. Still today, the messages exchanged among humanists about "tyranny" and "freedom" have as their referent the code of humanism promoted by Cola Montano. And the violence represented in these codified texts is sublimated in the suppression of every kind of "literature" which cannot be assimilated to this code.

It is important to understand the humanistic codification of classical narratives of political violence not as a screen isolating humanists from authentic political activity, but as a particular political strategy in its own right, a strategy of suppression. Just as the message of Thrasybulus prescribes the suppression of the prominent citizens of Corinth, so the message of humanists prescribes its own program of suppression, by isolating "literature" from other kinds of writing. What we call today "literature" is one among many complex and conflicting written codes which make up "reality." That this category of writing has been historically formed and determined by the cultural practice of a particular social group is undeniable. But the ways in which this socially specific practice has obstructed the study of "literature" in relation to other kinds of writing and other practices of literacy has yet to be fully acknowledged.

Cantimori intuited this work of suppression in his representation of Pier

Paolo Boscoli, humanist and tyrant-slayer, who, in the last moments before death, dimly perceived that the humanistic code was "not enough" for the understanding of "Renaissance life." And today, paleographers and scholars of the history of writing are striving to show how the humanistic "work of resemblance," by singling out for reproduction and transmission only the particular cultural products of humanistic activity, suppressed the plurality of manifestations of writing which flourished in this period of the birth of mass literacy.[16]

Today, it is particularly opportune to investigate these processes of suppression. As new kinds of non-humanistic literacy[17] are rapidly supplanting the humanistic code and establishing themselves as the new "literature," it becomes imperative for humanists to question the disciplinary boundaries we have set up between "literature" and the written documents of "real life." For the survival of the "humanities" may well depend on our recognizing and subverting the politics of suppression and domination that has kept us from making connections between the "literature" we study and the writing which determines, in a quotidian sense, our practices of interpretation. The possibility of making such connections depends upon the description of conditions under which we interpret and the practice of writing as a history of the conditions and qualities of being literate, as a history of establishing material relations between culture and everyday social and political life.

Notes

1 The conception of this essay depends upon the Althusserian notion of ideology and upon the use Hayden White makes of this notion (Althusser 1971: 162–77; White 1982a: 289).

2 See, in particular, Ricoeur's (1981) discussion of the debate between Gadamer and Habermas. In this essay, I consider the humanistic tradition, constituted as it is by particular kinds of texts, modes of transmission and cultural practices, as an exemplary field in which to practice the critique of ideology.

3 Here, *Aristobolo*, in all probability, represents a slightly altered memory of *Trasibolo*. Translations in this essay, unless otherwise indicated, are my own.

4 Some of these numerous recurrences include: Frontinus I.i.4; Diogenes Laertius I.vii.100; Ovid II.705–7; Cornelius Nepos I.1; Valerius Maximus VIII.iv. For a modern recurrence, see Kierkegaard, 1968.

5 Humanism is understood here in a technical sense as a term for the particular activities of editing, reproducing, studying, and transmitting classical texts performed by scholars of Greek and Latin literature in fifteenth- and sixteenth-century Italy. It is this author's understanding that our modernday humanistic activities are descended from those of the first humanists, even when we focus our hermeneutic attention on non-classical literature. See, in particular, Campana 1946 and Kristeller 1961.

6 Here I use the term "class" to refer to the group of individual subjects who understand the code of humanism to be "natural" and not specific to their own social practices (Althusser 1971: 170–7; White 1982a: 289). For a social history of fifteenth-century Florentine humanists, see Martines 1963.

7 For this relation between sign system and Peirce's concept of the "self-analyzing habit," see de Lauretis 1984: 158–86.
8 Petrucci cites 1456 here as a "probable" date.
9 What follows here is my paraphrase of Machiavelli's narrative (VII.34). For a modern historian's account of the same episode, see Ilardi 1972.
10 For a discussion of some of the consequences of such discriminations, see Valesio 1978 and White 1982b.
11 For biographies and analyses of the political/intellectual/methodological itinerary of this most eminent Italian historian, whose impatience with static abstractions led him to membership in the Italian Fascist Party in 1926, to a progressive detachment from fascism in the mid-1930s, to direct relations with the Italian Communist Party in 1940 (and membership in 1948), to a break with the PCI in 1956, see: Ciliberto 1977; Miccoli 1977; Garin 1974; "Cantimori" 1975.
12 See, for example, Cantimori 1941 and 1937a.
13 For alternative articulations of this point, see White who writes of "a code specific to the praxis of a given social group, stratum, or class" which is none the less treated as "natural" (1982a: 289); and Hjelmslev (1963) who writes of a linguistic form, in our case, the term "literature," "being projected on to the purport [in our case, the reading and writing practices of humanists], just as an open net casts its shadow down on an undivided surface" without acknowledging the division it produces (57).
14 Ginzburg (1982) and Davis (1983) have demonstrated the importance of interpreting the writing constituted, for example, by trial records.
15 Cantimori 1927: 251. For the notion that a politics is implicated in this sublimation, see White 1982b.
16 See the special issue of *Quaderni storici* 38 (1978) devoted to the field of "Literacy and Written Culture" and the *Notizie* of a "permanent seminar" devoted to this field of study (published by the Dipartimento di Scienze storiche at the University of Perugia); the journal *Scrittura e civilta*; Petrucci 1983; Graff 1983.
17 Two kinds of non-humanistic literacy immediately come to mind: literacy in computer languages and, more generally, literacy in the languages of electronic media.

References

Albertini, Rudolf von (1955) *Das Florentinische Staatsbewusstsein in Ubergang von der Republik zum Prinzipat*. Bern: A. Francke AG Verlag.

Althusser, Louis (1977) "Ideology and Ideological State Apparatuses (Notes towards an Investigation)," in *Lenin and Philosophy*. Trans. Ben Brewster. New York: Monthly Review Press: 127–86.

Aristotle (1885) *The Politics*. Trans. B. Jowett. Oxford: Clarendon.

Campana, Augusto (1946) "The Origin of the Word 'Humanist.'" *Journal of the Warburg and Courtauld Institutes* 9: 60–73.

"Cantimori, Delio" (1975) *Dizionario biografico degli italiani*. Rome: Instituto della Enciclopedia italiana.

Cantimori, Delio (1927) "Il caso del Boscoli e la vita del Rinascimento." *Giornale critico della filosofia italiana* 4: 241–55.

——— (1937a) "La vita come ricerca." *Giornale critico della filosofia italiana* 18: 356–70.

——— (1937b) "Rhetoric and Politics in Italian Humanism." Trans. Frances Yates. *Journal of the Warburg Institute* 1: 83–102.

——— (1941) "Appunti sulla propaganda." *Civilta fascista* 8: 37–56.
Cavriana, Filippo (1597) *Discorsi sopra i primi 5 libri di Tacito*. Fiorenza: Filippo Giunti.
Ciliberto, Michele (1977) *Intellettuali e fascismo: Saggio su Delio Cantimori*. Bari: De Donato.
Davis, Natalie Zemon (1983) *The Return of Martin Guerre*. Cambridge, Mass.: Harvard University Press.
della Robbia, Luca (1842) "Recitazione del caso di Pietro Paolo Boscoli e di Agostino Capponi . . . l'anno MDXIII." Ed. F. Polidori. *Archivio storico italiano* 1st ser. 1: 275–311.
de Lauretis, Teresa (1984) *Alice doesn't: Feminism, Semiotics, Cinema*. Bloomington: Indiana University Press.
Diogenes Laertius (1980) *Lives of the Eminent Philosophers*. Trans. R.D. Hicks. Cambridge, Mass.: Harvard University Press.
Eco, Umberto (1979) "The Semantics of Metaphor," in *The Role of the Reader*. Trans. Jon Snyder. Bloomington: Indiana University Press: 67–89.
Frontinus, *The Strategems* (1950) Trans. C.E. Bennett. Cambridge, Mass: Harvard University Press.
Garin, Eugenio (1974) "Delio Cantimori," in his *Intellettuali italiani del XX secolo*. Roma: Editori Riuniti: 171–213.
——— (1975) *Scienze e vita vivile nel Rinascimento italiano*. Bari: Laterza.
Ginzburg, Carlo (1982) *The Cheese and the Worms: The Cosmos of a Sixteenth-Century Miller*. Trans. John and Anne Tedeschi, Harmondsworth: Penguin.
Graff, Harvey J. (1983) "On Literacy in the Renaissance: Review and Reflections." *History of Education* 12: 69–85.
Habermas, Jurgen (1971) "Knowledge and Human Interests: A General Perspective," in his *Knowledge and Human Interests*. Trans. Jeremy J. Shapiro. Boston: Beacon Press.
Herodotus (1927) *Historiae*. 3rd edn. Ed. C. Hyde. Oxford: Clarendon.
Hjelmslev, Louis (1963) *Prolegomena to a Theory of Language*. Trans. Francis J. Whitfield. Madison: University of Wisconsin Press.
Ilardi, Vincent (1972) "The Assassination of Galeazzo Maria Sforza and the Reaction of Italian Diplomacy," in *Violence and Civil Disorder in Italian Cities*. Ed. Lauro Martines. Berkeley: University of California Press: 72–103.
Jakobson, Roman (1956) "Two Aspects of Language and Two Types of Aphasic Disturbances," in R. Jakobson and M. Halle (eds) *Fundamentals of Language*. The Hague: Mouton.
Jameson, Fredric (1981) *The Political Unconscious*. Ithaca, NY: Cornell University Press.
Kierkegaard, Soren (1968) *Fear and Trembling*. Trans. Walter Lowrie. Princeton, NJ: Princeton University Press.
Kristeller, Paul Oskar (1961) "Humanism and Scholasticism in the Italian Renaissance," in his *Renaissance Thought: The Classic, Scholastic, and Humanist Strains*. New York: Harper & Row: 92–119.
Livy (1955) *Ab urbe condita*. Ed. R.S. Conway and C.F. Walters. Oxford: Clarendon.
Machiavelli, Niccolo (1962) *Istorie fiorentine*. Ed. Franco Gaeta. Milano: Feltrinelli.
Martines, Lauro (1963) *The Social World of the Florentine Humanists 1390–1460*. London: Routledge & Kegan Paul.
Miccoli, Giovanni (1977) *Delio Cantimori: La ricerca di una nuova critica storiografica*. Torino: Elnaudi.
Nepos, Cornelius (1929) *Liber de excellentibus ducibus exterarum gentium*. Trans. J.C. Rolfe. Cambridge, Mass.: Harvard University Press.

Ovid (1959) *Fasti*. Trans. J.G. Frazer. Cambridge, Mass: Harvard University Press.
Petrucci, Armando (1983) ''Le biblioteche antiche.'' *Letteratura italiana* 2: *Produzione e consumo*. Torino: Elnaudi: 527–54.
——— (1983) *Scrittura e popolo nella Roma Barocca (1585–1721)* Roma: Quasar.
Ricoeur, Paul (1981) ''Hermeneutics and the critique of ideology,'' *Hermeneutics and the Human Sciences*. Ed. and trans. John B. Thompson. Cambridge: Cambridge University Press: 63–100.
——— (1977) ''The work of resemblance.'' *The Rule of Metaphor*. Trans. R. Czerny with K. McLaughlin and J. Costello, SJ. Toronto: University of Toronto Press: 173–215.
Shakespeare, William (1962) *The Tragedy of Richard the Second*. Ed. Louis B. Wright. New York: Simon & Schuster.
Valerius Maximus (1539) *Dei detti et fatti memorabili*. Trans. Giorgio Dati. Roma: Antonio Blado d'Asola.
Valesio, Paolo (1978) ''*Pax Italiae* and the Literature of Politics.'' *Yale Italian Studies* 2: 143–68.
White, Hayden (1982a) ''Method and Ideology in Intellectual History,'' in Dominick LaCapra and Steven L. Kaplan (eds) *Modern European Intellectual History: Reappraisals and New Perspectives*. Ithaca, NY: Cornell University Press: 280–310.
——— (1982b) ''The Politics of Historical Interpretation: Discipline and De-Sublimation.'' *Critical Inquiry* 9: 113–37.

2

"Drunk with the Cup of Liberty"

Robin Hood, the carnivalesque, and the rhetoric of violence in early modern England

Peter Stallybrass

> [Among] the lower class of people [there is] such brutality and insolence, such debauchery and extravagance, such idleness, irreligion, cursing and swearing, and contempt of all rule and authority. . . . Our people are drunk with the cup of liberty. (Josiah Tucker, Dean of Gloucester, 1745)

Matthew Arnold's famous opposition between Culture and Anarchy can, perhaps, be seen as one way of constructing an opposition between rhetoric and violence. Rhetoric, at least the "legitimate" rhetoric of the classic text, becomes a machine to overcome time, a transhistorical reason which can rebuke and exorcize the specter of anarchy. For Arnold, that "Anarchy" was embodied in the crowd which tore down the railings of Hyde Park, thus erasing the boundaries between order and disorder, between a culture enclosed like the park and the limitless horizons of chaos (Arnold 1932). In the twentieth century, though, Arnold's "Culture" was to be rethought as repression: rhetoric and the regularities of language and discourse were no less than the structure of the dominant social order. For many critics of modernism, linguistic transgression became a privileged form of politics, since it violated the very terrain on which more "conventional" political activity was always already situated. Kristeva, for instance, suggests that "there is no equivalence, but rather identity between challenging official linguistic codes and challenging official law." Carnivalesque discourse, she argues, in violating "the laws of a language censored by grammar and semantics" is, at the same time, "a social and political protest" (Kristeva 1980: 65).

Curiously, although Kristeva explicitly defines carnivalesque dialogue as "*transgression giving itself a law*" (1980: 71), her argument tends to slide toward the Arnoldian oppositions in which order, rhetoric, and reason confront violence, linguistic transgression, and unreason. Of course, Kristeva's evaluation of these polarities is more or less the opposite of Arnold's. Yet, in so far as she conceptualizes linguistic laws and violence as opposites, she defines transgression as an operation from the margins, always bordering on silence. At the same time, in her appropriation of Bakhtin's concept of the carnivalesque, she tends to reduce carnival to the fracturing of discourse, thus privileging the practices of literary signification at the expense of the significations of social practice. The carnivalesque becomes identified with the linguistic transgressions of the individual artist (Artaud or Céline) who confronts the monolithic orthodoxies of the social – that is, individual "anarchy" subverts collective "culture."

I would argue that we should analyze the carnivalesque as a set of rhetorical practices *within* the social, a set which includes, but is by no means limited to, linguistic devices. A tentative morphology of the carnivalesque might include the following elements:

1. The replacement of fast by feast.
2. Transgression of spatial barriers. The marketplace as the locus of public life encroaches upon the privacy of houses. Locked doors are broken down; windows are smashed (Thomas 1976: 16, 27). At the same time, the boundaries of the village or town may become permeable, the inhabitants processing out into the fields, woods, and forests.
3. Transgression of bodily barriers. The carnivalesque emphasizes those parts which rupture the "opaque surface" of the classical body: "the open mouth, the genital organs, the breasts, the potbelly, the nose" (Bakhtin 1968: 26). The head is subordinated to "the lower bodily stratum"; the open mouth and the nose are foregrounded because they are equated with the anus and the phallus.
4. The inversion of hierarchy. The servant rules his master, the child rules his parents, the wife rules her husband.
5. The degrading of the sacred.
6. The transgression of the linguistic hierarchy. The languages of billingsgate (oaths, curses, obscenities) are privileged over "correct" speech (Bakhtin 1968: 15–17).

But a morphology of the carnivalesque cannot assign a formal meaning in advance. Indeed, the collapse of meaning into morphology repeats the subsumption of "coutumes populaires" by "discours bourgeois," through which the folklorist, gazing upon past forms, "etudera l'acts vidé de son contenu concret, dépouillé de son contexte matériel" (Bonnain-Moerdyk 1977: 382). Carnivalesque meaning, never a simple given, is made and remade in the contested domain of social practices. It is both the forms and the uses of a carnivalesque rhetoric of violence which we shall analyze in the production and reproduction of Robin Hood in early modern England.

Robin Hood and the carnivalesque

Carnivalesque discourse permeates even the earliest ballads of Robin Hood:

he transgresses spatial boundaries, trespassing upon the king's forests; he inverts social hierarchy, preying upon the dominant classes; he degrades the sacred, humiliating the clergy; he replaces the fast of peasant poverty with the feast of poached deer. But the outlaw owes the perpetuation of his legends and his widespread fame less to the ballads than to his incorporation into the carnivalesque festivities of the May-games, with their plays, pageants, and morris dances. A medieval stained glass window represents eleven morris dancers, including Maid Marion, a friar, a hobbyhorse, and a fool, but none of the figures can be identified with Robin Hood (Simeone 1951: 268). It was only in the late fifteenth and early sixteenth century (long after the outlaw was established in the ballad tradition) that Robin Hood became associated with the May games and with the morris dance where, for the first time, his name was linked with Maid Marion and Friar Tuck. In 1473, we hear of a servant employed "to pleye Seynt Jorge and Robyn Hod, and the Shryff off Nottyngham" (Dobson and Taylor 1976: 38); in 1505, "Robyn Hod of Handley and his company" acted in the May festivities at Reading St Lawrence (Chambers 1903: 176); in 1509, the accounts of Kingston-on-Thames record money spent on the costumes of Robin Hood, Little John, Maid Marion, and a friar (Simeone 1951: 271); and the accounts of Croscombe note money given to Robin Hood for the "coming years of stewardship" (Simeone 1951: 272).

Nor was the carnivalesque play of Robin Hood always confined to the *licensed* misrule of May. In his *Chronicle*, Robert Fabyan records arrests for treason in 1502, adding that "Also thys yere about Midsomer was taken a felowe, whych had renued many of Robin Hodes pagentes, which named himself Grenelef." We do not know what subversions Grenelef's "pagenetes" conceal, but "Reynold Grenelefe" was the name which Little John had assumed in the ballad, "A Gest of Robyn Hode," when he lured the Sheriff of Nottingham into the forest (Nelson 1973: 28–31). In Scotland, too, the boundaries between May play and the playing out of ritual subversion were unclear. The Scottish Parliament decreed in 1555 that anyone impersonating Robin Hood would be banished and anyone electing a Robin Hood would be deprived of his or her freedom for five years. But in 1561, a crowd elected a tailor as Robin Hood, naming him "Lord of Inobedience," and "efter the auld wikit manner of Robene Hude" (Mill 1927: 221–3) they

> seized on the city gates . . . and one of the ringleaders being condemned by the city magistrates to be hanged, the mob forced open the jail, set at liberty the criminal and all the prisoners, and broke in pieces the gibbet erected at the gates for executing the malefactor. (Simeone 1951: 273)

The magistrates themselves were then imprisoned until they published a proclamation pardoning the rioters. In this instance, the "law" of carnivalesque play triumphed over statutory law: the signs of the legal apparatus (the gibbet, the prisons) were destroyed or transgressed. But four years later, several citizens were disenfranchised for breaking the 1555 decree which prohibited any ''tumult, scism, or conventione'' with ''taburne playing, or pype, or fedill . . . in chusing of Robin Huid, Litill Johnne, Abbot of Resoune, Queyne of Maii'' (Chambers 1903: 181; Simeone 1951: 273).

As the outlaw was incorporated in the May games, so the Robin Hood ballads of the seventeenth century included new elements drawn from carnival. In ''Robin Hood's Birth, Breeding, Valor and Marriage,'' after feasting on hot venison, warden pies, ''cream clouted, with honey-combs plenty,'' the outlaw carries his poached deer into Titbury where ''the bagpipes bated the bull'' and ''some were a bull-back, some dancing a morris'' (Child 1965: 216–17). In ''Robin Hood and the Bishop,'' published between 1620 and 1655, we can analyze in more detail how carnivalesque discourse structures a particular text (Child 1965: 191–3). The ballad is composed of seven main actions:

a. Robin Hood, ranging alone in the forest, sees a bishop and his company and flees, fearing capture;
b. He changes clothes with an old woman;
c. The bishop arrests the old woman, believing her to be Robin Hood;
d. The bishop is captured by Robin Hood, who has rejoined his men;
e. Robin Hood takes £500 from the bishop;
f. Little John forces the bishop to say mass and pray for Robin Hood;
g. The bishop is put on a horse back to front and released.

The ballad depends upon both the inversion of hierarchy and the profanation of the sacred. The thematic inversion is paralleled by a linguistic inversion of the first stanza: ''I'le tell you how Robin Hood *served* the Bishop. . . .'' It is the required deference of ''service'' of the subordinated to their religious and social superiors that Robin Hood subverts.

Robin Hood's triumph over the bishop, though, is made possible by a prior carnivalesque inversion. For although the outlaw is ''stout,'' he ''turned him about'' on seeing the bishop and his company,

> And to an old wife, for to save his life,
> He loud began for to cry.

In the narrative, the fulfillment of Robin Hood's request to exchange clothes with the old woman is motivated by the statement that he has previously given her ''shoes and hose.'' But metaphorically the old woman's transformation into Robin Hood enacts the carnivalesque inversion which gives power to the ''unruly'' woman. In fact, the old woman participates in a series of inversions: she is dressed as a male outlaw; she is set upon a milk-white steed (commonly associated with virginity and religious virtue) whilst the bishop rides a ''dapple-grey;'' she derides the bishop. Robin Hood undergoes an opposite series of transformations. Although ''stout,'' he flees, begs mercy of an old woman, and assumes her clothes which make him look "like an old witch." He exchanges male for female, bow and arrows for "spindle and twine," a "mantle of green" for a "coat of grey," youth for age.

These transformations have disturbed those critics who wish to find in Robin Hood a romance hero. But they are, I believe, explicable as signs in a specific social process: the charivari. The charivari (in England usually called the rough ride, rough music, or skimmington) was a noisy procession in which the victim, or an effigy of the victim, was paraded through the streets and ridiculed by the community (Sharp 1980: 104–5; Bonnain-Moerdyk 1977; Burguière 1980; Tilly 1980). The ballad's narrative is resolved by two charivaris. The first is formed by the transformation of the bishop's triumphal procession ("And for joy he had got Robin Hood,/He went laughing all the way") into a mock ceremony of which he is the victim:

> "Why, who art thou," the Bishop he said,
> "Which I have here with me?"
> "Why, I am an old woman, thou cuckoldly bishop;
> Lift up my leg and see."

The second charivari completes the ballad when the bishop, having been forced to say mass and pray for Robin Hood, is put on his horse back to front, the horse's "tail within his hand." (Another version of this can be found in "A True Tale of Robin Hood" (1632) where the bishop is made to ride "with his face towards [the horse's] arse" (Child 1965: 2,281)).

Riding backwards is a rite of humiliation with a long history. In 998, the antipope, Johannes Philagatos, was mutilated and then made "to ride backwards on an ass wearing his vestments turned inside out" (Mellinkoff 1973: 154–5). Wherever the custom originated, though, it was used in early modern England as a popular form of social regulation and protest. Common targets for this sort of skimmington were the husband who did not control his wife, the cuckold, the adulterer, and the "unruly" woman. John Stow, for instance, in his *Survey of London*, describes a riding which he saw in his youth after John Atwood, a draper, discovered that his wife was having an affair with a chantry priest. Any one of the three would have been a suitable target for a charivari, but, after Atwood was reconciled to his wife, it was the priest who was "apprehended and committed":

> he was on three Market dayes conveyed through the high streete and Markets of the Citie with a Paper on his head, wherein was written his trespasse: The first day hee rode in a Carry, the second on a horse, his face to the horse taile, the third, led betwixt twaine, and every day rung with Basons, and proclamations made of his fact at every turning of the street. (Stow 1908: 190)

The ritual procedures are typical: the public procession through the high streets and markets; riding backwards; the raucous noise of "Basons" being beaten. (In 1704, Thiers defined charivari in terms of noise alone: ''To make noise with drums, firearms, bells, platters, plates, pots, skillets, casseroles, and cauldrons; to hoot, whistle, jeer, and cry in the streets; in a word, to make what is called charivari'' (Burguière 1980: 931).) We should note that Stow states that the priest was ''apprehended and committed'': his three-day punishment was, indeed, a popular version of the *official* penance that the church courts had the statutory right to enforce.

The charivari in early modern England was upheld by the manor courts as a quasi-legal means of correcting ''domestic crimes.'' In East and West Enborne, an ''incontinent'' widow lost her rights to free bench (which entitled her to the possession of a portion of her deceased husband's lands during her life), unless she rode backwards upon a black ram to the next manor court, reciting:

> Here I am,
> Riding upon a Black Ram,
> Like a Whore as I am;
> And for my Crincum Crancum,
> Have lost my Bincum Bancum;
> And for my Tail's Game,
> Am brought to this worldly Shame,
> Therefore good Mr. Steward let me have my lands again.
> (Thompson 1976: 335, 349–58; Mellinkoff 1973: 161)

Here, popular ritual is incorporated into the workings of the legal apparatus.

In ''Robin Hood and the Bishop,'' though, the symbolic system of the rough ride is manipulated by an outlaw who is described by the bishop as ''That traitor Robin Hood,'' and it is directed not against the unruly personal conduct of a subordinate, but against the enforcers of political orthodoxy. The old woman is the pivot of the carnivalesque in the ballad. She can herself be seen as a typical example of the charivari's victim. Although she is called the ''old *wife*'' twice, the first time the signifier is phonologically motivated by the rhyme with ''life.'' She appears to be an old woman living alone throughout the ballad and may be taken to be a widow (thus reconciling the term ''wife'' and her single status). Of all women, widows were most subject to charges of lechery, witchcraft, scolding, and other forms of unruliness. But in the ballad, the accusation of witchcraft is turned into a jest against Robin Hood, who is perceived by Little John as an ''old witch.'' And the old woman's unruliness, far from being penalized, becomes a symbolic weapon against another common type of charivari victim – the cuckold:

> Why I am an old woman, thou cuckoldly bishop;
> Lift up my leg and see.

The gulling of the bishop is inscribed as his cuckolding by a crone. The sign of her gender (''Lift up my leg and see'') is simultaneously the mark of his impotence.

This subversion of class and gender hierarchy is underwritten by an implicit appropriation of the moral economy (see Thompson 1971). Although an old woman could become the target for a skimmington, she was also an object of *charity*, a charity preached, and in theory practiced, by the Church. But in the text, the bishop is ''proud'' and, at least towards outlaws, has ''No mercy,'' whereas the ''traitor,'' Robin Hood, protects the poor:

> ''If thou be Robin Hood,'' said the old wife,
> ''As thou dost seem to be,
> I'le for thee provide, and thee I will hide
> From the Bishop and his company.
>
> For I well remember, one Saturday night
> Thou bought me both shoos and hose;
> Therefore I'le provide thy person to hide,
> And keep thee from thy foes.''

The symbolic system of carnival is used, then, to legitimate popular justice against the official ideological and legal apparatus which claims to have a monopoly of justice. The outlaw becomes the enforcer of popular law.

Carnival and the interrogation of rule

''Robin Hood and the Bishop'' celebrates the triumph of ''dirt'' or ''matter out of place,'' to use Mary Douglas's term (1978: 35). The boundaries of gender, class, status, the body, the sacred, and the cultural are transgressed: female mocks male; the oppressed deride the oppressor; the human is subordinated to the animal; the head is turned toward the anus; prayers are said for the outlaw; mass is parodied in the wilderness. Of course, such carnivalesque inversions could be used to reaffirm local norms of domestic hierarchy, as we have already observed. Indeed, Charles Tilly (1980) argues that *all* charivaris before the French Revolution ''remained within the limits set by domestic morality'' (1980: 9), and he claims that the political charivari was an invention of the late eighteenth century and was developed in the struggles of the nineteenth century. On the contrary, I would argue that elements of carnival and charivari were central to the symbolic repertoire of political subversion in early modern Europe. I want to examine, in greater depth, three of these elements which we have already touched on in ''Robin Hood and the Bishop'': the uses of dirt or excrement; gender-inversion and transvestism; the ''liberties'' of the forest.

The uses of dirt

In riding backwards, the lower bodily stratum, to use Bakhtin's term (1968: 20–1), triumphs over a hierarchy which privileges the head. But carnivalesque dirt, like carnivalesque laughter, is radically ambivalent, being both a means of derision and debasement and a cause for celebration. In "Robin Hood and the Bishop," derision is dominant, but in the play of "Robin Hood and the Friar" (*c.* 1560) ("to be played in Maye games very pleasante and full of pastyme"), dirt intersects with the celebration of sexuality (Dobson 1976: 208). The play stages a battle of wits between Robin Hood and Friar Tuck in which Robin's anticlericalism is foregrounded ("He never loved fryer nor none of freiers kin" (Dobson and Taylor 1976: 211)). In the conclusion, the "lowsy frere" is drafted into the service of Robin Hood, and Robin makes him the chaplain of "a lady free" (presumably Maid Marion). The final speech is given to the friar who celebrates his "service" of the lady:

> Here is an huckle duckle
> An inch above the buckle.
> She is a trul of trust,
> To serve a frier at his lust,
> A prycker, a prauncer, a terer of sheses [sheets],
> A wagger of ballockes when other men slepes
> Go home, ye knaves, and lay crabbes in the fyr,
> For my lady and I wil daunce in the myre for veri pure joye.

These lines are omitted from Child's edition of the play on the grounds that they are not relevant to the Robin Hood tradition (Child 1965: 128). But they are crucial in analyzing the workings of carnivalesque discourse. Dirt is used in the play both to degrade the Church (the friar is literally brought down to "the myre") and to celebrate the liberties of carnival (in "the myre" one dances "for veri pure joye").

The connection between dirt and the subversion of the Church was inscribed within the symbolic practices of carnival. In Saxony in June 1524, a mock bishop was dressed in a straw cloak and carried a fish basket as his mitre. His canopy was a filthy cloth, the holy water vessel was an old fish kettle, candles were dung forks. "Sacred" relics (a horse's head, two horse-legs, the jawbone of a cow) were placed on a dung carrier and covered with old bits of fur and dung. In Alsace, at the feast of the Magi in 1525, youths elected a mock king and, preceded by a piper and a drummer, processed to the ecclesiastical foundation of St Leonard, demanding a "gift for their king." They were refused, but they returned in April, and destroyed images, altars, and religious books, satirized church ceremonies in a *Narrenspiel*, and defecated on the altar. On *Fastnacht* 1521 in Wittenberg, a mock pope was hunted through the streets and pelted with dung (Scribner 1978: 306, 307,

304). Dung was even incorporated into the anti-rites *within* the church at the Feast of Fools: ''During the solemn service sung by the bishop-elect, excrement was used instead of incense. After the service the clergy rode in carts loaded with dung; they drove through the streets tossing it at the crowd'' (Bakhtin 1968: 147). In the parody of ecclesiastical rule, the spiritual is displaced by the material, the top by the bottom in a ritual in which dirt simultaneously degrades the sacred and celebrates the ''base'' – that is, what is both physically and socially *low*.

Dirt was also used in the symbolic subversion of the state. We can trace some of the connections between the destruction of enclosures, dirt (breaking the enclosure of the body) and popular revolt in Kett's rebellion of 1549. In July, crowds gathered for the Wymondham Game, and the festivities lasted from Saturday, July 6, to the following Monday. On Tuesday, some of the celebrants began to pull down enclosures made by the gentry (Beer 1982: 83). Some 16,000 commoners then gathered under the leadership of Robert Kett, marched on Norwich and, by July 23, had taken the city. But how was the city captured? According to the earliest accounts, the defenders were overpowered not by weapons but by the naked arses and obscene gestures of ''vagabond boyes'' who ''brychles and bear arssyde came emong the thickett of the arrows . . . and did therewith most shamefully turne up theyr bare tayles agenst those which did shoote, which soe dysmayd the archers that it tooke theyr hart from them'' (Beer 1976: 87; Fletcher 1973: 66).

This is Sotherton's account, and Neville also describes the ''odious and inhumane villainy'' of a boy who pulled down his hose and, turning his bottom to the defenders, ''with an horrible noise and outcry, filling the air . . . did that, which a chaste tongue shameth to speak, much more a sober man to write'' (Beer 1982: 100). Holinshed recounts a similar incident when the king's herald was sent to the rebels' camp to offer pardon. The herald was greeted by ''a vile boie'' who ''turned up his bare taile to him, with words as unseemlie as his gesture was filthie'': outraged, the herald's attendants killed the boy (Holinshed 1808: 979). Of course, Sotherton, Neville, and Holinshed were all concerned to vilify the rebels. It is not necessary, though, to accept Sotherton's claim that it was the ''bare tayles'' of ''vagabond boyes'' which won the battle to understand the rhetorical force of the gesture to a social elite which depended upon deference even more than upon military strength. For the dominant classes, the relations between ''polite,'' ''police,'' and ''politics'' were more than a question of etymology. In the rebel camp, though, a boy's ''baire taile'' subordinated top to bottom, age to youth, monarchy to commoner, and courtliness to dirt. Similarly, in ''A True Tale of Robin Hood,'' an outlaw made a dignitary ''face towards [the horse's] arse'' (Child 1965: 228).

Gender-inversion and transvestism

In carnival, gender, like dirt, provided a concrete logic through which the subversion of deference and hierarchy could be rehearsed. As we have noted, the narrative of "Robin Hood and the Bishop" hinges upon both transvestism and the inversion of gender hierarchy. The Robin Hood who is arrested is played by an old woman, and the revelation of her identity leads to a carnivalesque triumph over the "cuckoldly bishop." Certainly, "Robin Hood and the Bishop" is atypical amongst the outlaw ballads, where gender is rarely foregrounded. But apart from the early sixteenth century, as we have seen, Robin Hood was incorporated into the May games, where he became associated with Maid Marion and Friar Tuck. In the play of "Robin and the Friar," the "lady free" is still paired with the Friar, not with Robin Hood, and the play, "verye proper to be played in Maye games," probably ended in a morris dance in which Maid Marion was usually played by a man or boy. Only in the high literary tradition did Marion become a chaste damsel; in the popular tradition, she was "a smurkynge wench" and "none of those coy dames" (Dobson and Taylor 1976: 42), exchanging lewd jokes with the Fool who accompanied her.

Long before her connection with Robin Hood, Marion had been an important figure in the May games, a privileged time for gender-inversion. In ancient Rome, May had been Flora's month, when women were supposedly most powerful, and proverbially a May wife would rule over her husband. Natalie Davis (1978) has collected evidence of the unruly woman of May in early modern Europe. In Franche-Comté, women could duck their husbands or make them ride on an ass, and they could feast and dance without their husband's permission. In Dijon, Mère Folle and her Infanterie conducted a charivari against a man who had beaten his wife (Davis 1978: 170). No doubt, the Marion of the May games also owed something to other festivities: Hock Tuesday, when women bound up men and only released them for a ransom; the Feast of Fools, when clerks, dressed as women or panders, danced in the choir singing bawdy songs or ate black pudding at the altar (Chambers 1903: 293). And, in so far as Marion was confined to a particular festival, she probably remained largely a figure of *licensed* misrule.

But the unruly woman (whether played by an actual woman or by a transvestite man) could be used to emblematize the revolt of the powerless in more radically subversive ways. On *Fastnacht* 1543 in Hildesheim, women ridiculed and abused a group dressed as monks and clergy before expelling them from the city (Scribner 1978: 309). As in "Robin Hood and the Bishop," the subordination of women is inverted as women deride the male hierarchy of the Church. Women were equally prominent in challenging local government. In Maldon, Essex, there were two major food riots in 1629. The first, in March, was composed entirely of women who boarded a ship and forced the crew to fill their aprons and bonnets with rye. The dearth continued and unemployment amongst clothworkers also rose steadily. In May, the privileged month of gender-inversion, there was a second riot of male weavers who appropriated large quantities of grain. The riot was organized and led by Ann Carter, who had already been questioned for her part in the March riot. Carter had been in trouble with the authorities before. In 1622, she had attacked one of the town's two chief magistrates "calling him a bloud sucker . . . and many other unseemly tearmes" (Walter 1980: 58). The following year, she abused the bailiff who came to question her about her absence from church. In May 1629, she assumed the title of "Captain," saying to her followers: "Come, my brave lads of Maldon, I will be your leader for we will not starve" (Walter 1980: 72). Two centuries later, rioting weavers were again led by female figures, "General Ludd's Wives," in an attack upon steam looms in Stockport (Thompson 1963: 567). This time, though, the leaders were men dressed as women. The unruly "Marion" of carnival was used to legitimate political action by the powerless.

It was not in England, though, but in France that the name of Marion was to become indissolubly linked to revolution. As Maurice Agulhon (1981) has shown, Marianne was the name given to the French Republic by reactionaries. The reasons for such a christening are obscure, but it may be noted that in the medieval French *fêtes du mai*, the shepherdess, Marion, presided with her lover, Robin, over rural pursuits, whilst various aristocrats attempted to seduce her (Chambers 1903: 171–2). Was the name "Marianne" a way of showing patrician contempt for a peasant and plebeian republic? However the republic came to receive its name, though, the revolutionary figure of Marianne was constructed within the symbolic language of the carnivalesque unruly woman. In Delacroix's famous painting, she is Liberty, bare-breasted, a flag in one hand, a rifle in the other, upon her head the Phrygian cap which had been the emblem of emancipation for slaves in ancient Rome. In the *Iambes* of Auguste Barbier,

> She is a strong woman with thrusting breasts,
> A harsh voice and a hard charm,
> Who, with her bronzed skin and her flashing eye . . .
> Takes her lovers only from among the people. (Agulhon 1981: 40)

Curiously enough, the iconography of the radical Marianne, as it is embodied, for instance, in Rude's sculpture of *La Marseillaise* on the Arc de Triomphe, is an appropriation of the classical iconography of Anarchy, "'a shrew' whose stance betokens fury and who holds in her hands 'a dagger and a flaming torch, alluding to the crimes that she occasions'" (Agulhon 1981: 13). Patrician demonization was transformed into plebeian celebration.

Like the carnivalesque unruly woman, Marianne existed simultaneously as abstract emblem and as living embodiment. In November 1793, she was the actress who stood upon the altar of Notre-Dame-de-Paris, transforming the Church of Mary and orthodoxy into the space of Marianne and revolution. (In 1848, a reactionary Breton doctor would complain "she [the republic] takes your daughters and, on the very altars where in the past you used to come to worship God, she exposes the latter, half-naked, to the eyes of libertines" (Agulhon 1981: 27, 29).) In 1830 she was a young girl, "a latter-day Joan of Arc," who fought on the barricades and captured a piece of artillery before being transformed into the Liberty who, seated in a chariot, made her triumphal progress through the streets of Paris "amid cheers, cries of joy, and singing." In 1871, she was a "former governess marching in the front line, vigorously brandishing the staff of a red flag" or, in Bollène, she was a poor servant, dressed in red and wearing a Phrygian cap, who was paraded through the streets "like a live image of the radical Republic" to the accompaniment of the *Marseillaise* (Agulhon 1981: 42, 140, 151). (Three centuries before, in 1580, the woman who led La Mur's resistance to the Duke of Mayenne and his royal army had been called "Red Petticoat" (Le Roy Ladurie 1981: 247).)

In the French Revolution, popular protest was associated with the figure of an actual woman, but rural enclosure riots were often led by transvestites. If, in the May games, an unruly Maid Marion collaborated with a notorious outlaw who trespassed upon the King's forests, so, inversely, the symbolic transgression of enclosures was accompanied by carnivalesque transvestism. In Ireland between 1760 and 1770, the Whiteboys, dressed in white frocks and with blackened faces, destroyed enclosures, harassed landowners, and vowed "to restore the ancient commons and redress other grievances" (Davis 1978: 181). In the Pyrenees, peasants wearing white shirts like women's clothes and women's hats waged the "War of the Demoiselles" in defense of their right to firewood and pasturage in the forest (Davis 1978: 178–9). The riots in Braydon Forest, Wiltshire, of May and June 1631 were led by three men dressed as women who called themselves "Lady Skimmington" (Sharp 1980: 98, 100).

But the image of the unruly woman was ambivalent. If it could be used as a symbolic form through which the powerless contested the social order, it could also be used to reaffirm what Marvell called "Nature's Law" – that is, deference (Marvell 1971: 157). In 1635, the three "Lady Skimmingtons" of the Braydon riots were fined £500 each and then pilloried, wearing women's clothes (Sharp 1980: 108). Their carnivalesque disguise was transformed by the new context into a mark of humiliation. This ambivalence concerning female misrule is suggested by a group of women enclosure rioters who, arraigned in the Star Chamber, argued that "women were lawlesse, and not subject to the lawes of the realme as others are but might . . . offend without drede or punishment of the law" (Walter 1980: 63). For the court, it was precisely because the women were "lawlesse" that they were on trial. But from the perspective of carnivalesque misrule, women were liberated from the law's metaphysical enclosure in which the "rich and mighty . . . ride over the common people" and "hedge in some to be heirs of life and hedge out others" (Thomas 1966: 62–3). And in the overthrow of legal and agricultural enclosures, the unruly woman also interrogated the social enclosures through which gender was constructed. Using the rhetoric of carnival, the old woman could deride the bishop, her sexual, spiritual, and social "superior."

The liberties of the forest

May was the month of Maid Marion. It was also the month when the regulations of village and urban life gave way to the liberties of the forest. "Against May, Whitsonday, or other time," wrote Stubbes in *The Anatomie of Abuses*, "all the yung men and maides, olde men and wives, run gadding over night to the woods, groves, hils, and mountains, where they spend all the night in pleasant pastimes" (1877: 149). Or, in Spenser's words, "the mery moneth of May" is the time when "to the greene Wood they speeden hem all" (Spenser 1912: 436). It was the time, according to the Mayor of Coventry in 1480, when "the people of every great city . . . do harm to divers lords and gentles having woods and groves nigh to such cities by taking of boughs and trees" (Phythian-Adams 1979: 114–15). The tales of Robin Hood are also commonly set in May. In May, the "merriest month," people gathered in the "feyre foreste" under "the grene wode tree"; in May, the liberties of the outlaw were celebrated.

The "liberty" of the forest was more than a festive fiction, though. The forest was a gathering place for masterless men seeking a meagre subsistence, for cottagers and squatters who sometimes enjoyed freedom of tenure as well as for outlaws and religious dissenters (Hill 1975: 43–56). In the forest, the surveillance of the clergy and the gentry was at its weakest. In 1610, James I complained to the Commons that the forests were "nurseries and receptacles of thieves, rogues, and beggars" (Hill 1975: 51); proposals for "improving" waste lands, submitted to Parliament in 1653, stated that "those vast, wild, wide forests . . . do administer liberty and opportunity unto villainous minds to perpetuate and commit their wicked and vicious actions" (Thirsk 1972: 136). John Aubrey thought of forest dwellers as "mean people [who] live lawless, nobody to govern them, they care for nobody, having no dependence on anybody" (Wrightson 1982: 172). But the distinction between the lawless and the lawful forest dweller was usually hard to draw. Laborers, husbandmen, and artisans were dependent upon the use of commons and forest wastes for survival. Pigs, cattle, and sheep could be grazed there, timber and firewood collected, rabbits, hares, game birds, and deer poached. If poaching was definitely illegal, grazing and the collection of wood were both dependent upon customary rights and subject to the competing claims of landowners and the poor.

The forest was thus the terrain on which the very definition of ''liberties'' was fought out. From the perspective of the large landowners (and above all, the king himself), forest law was a means of both maximizing profits and asserting status. James I and Charles I parcelled out ''forests, chases, and wild and waste grounds'' to ''the private benefit of themselves and the general good of the commonwealth,'' claimed John Smith of Nibley (Thirsk 1972: 123). In the Forest of Dean, Gloucestershire, for instance, James I set up new enclosures and tried to raise money by mining iron ore and by disafforestation, a policy which Charles I pursued even more vigorously (Sharp 1980: 175–219). In Rockingham Forest, Robert Cecil planned to sell off woods, reduce the number of deer, and make new cattle pastures despite the opposition of ''the base sort'' (Sharp 1980: 171). At the same time, there were frequent attempts to challenge the customary rights of forest dwellers. A report of 1609 stated:

> At this time the waste soyle in all forests is moste extreamly pesterred and surcharged with all manner of beastes and cattell as well as with sheepe goates and swyne being beastes not commonable within a forrest by lawe. . . . [New] erections are also inhabited with lewd poore people such as doe live upon the spoyle and destruction of the kinges woodes and deere. (Sharp 1980: 172)

But the game laws were used not only for economic reasons but also to maintain the royal prerogative and as a means of social stratification. William the Conqueror had asserted his right to establish ''forests'' wherever he chose, and no one had the right to pursue game there without royal permission. This game law had fallen into disuse by the sixteenth century, but James I introduced new statutes governing hunting which required freeholders to have an income of £40 a year, persons with life estates £80 a year, and others £400 in personal property before they could kill game (Kirby 1931: 239, 241). New foresters, warreners, warders, parkers, and gamekeepers were appointed to enforce these laws and the Court of the Star Chamber was kept busy with prosecutions for illegal hunting (Sharp 1980: 171). The ''play'' of the gentry was increasingly symbolized by the pursuit of game, a pursuit which, in turn, transformed the countryside into game parks – the emblems of gentry prestige and privilege.

Forest law, though, met fierce opposition throughout the early seventeenth century. Enclosures were destroyed and deer killed in Gillingham Forest (1628), Leicester Forest (1628), Braydon Forest (1631); and in Waltham Forest, between April 1642 and February 1643, large crowds killed more than 400 of the king's deer (Sharp 1980: 88, 90, 108, 223). Even more strikingly, there were no deer left in Windsor Great Park by 1640, whilst in Windsor Forest as a whole there were 2,689 fallow deer in 1607 and only

203 in 1697, despite restocking after the Restoration (Thompson 1975: 56). Undoubtedly, attacks upon forest enclosures and deer were economically motivated. Enclosure curtailed the rights of pasturage and the gathering of wood; deer were both voracious and destructive. In 1640, a Berkshire petition was presented against "the innumberable increase of deer, which . . . will neither leave foode nor roome for any other creature in the forest" (Thompson 1975: 55).

Forest riots, though, were not outbreaks of "spontaneous" violence. On the contrary, they tended to be acted out through the symbolic practices of the carnivalesque. The 1631 riot in the Forest of Dean, for instance, can be seen as a charivari. A mock army, carrying weapons and led by "two drummers, two coulers and one fife" buried an effigy of Sir Giles Mompesson (the notorious monopolist caricatured as Sir Giles Overreach in Massinger's *A New Way to Pay Old Debts*) in an ore pit dug on ground which his men had enclosed and disafforested. The "army" then assembled outside an agent's house and announced that "if this deponent would make the like worke against May Day next they would be ready to doe him the like service" (Wrightson 1982: 177; Sharp 1980: 95–6). The ritualistic nature of the riot was foregrounded by the "military" procession, the burying of the effigy, the music, the legalistic reading of a charge against the "deponent" and the threatened punishment which was legitimated as May misrule. Nor can poaching, at least on a large scale, be reduced to a simple economic reflex. As E.P. Thompson says in relation to the Blacks (early-eighteenth-century deer poachers who blackened their faces), hunting was "retributive in character and concerned less with venison as such than with deer as symbols (and as agents) of an authority which threatened their economy, their crops, and their customary rights" (1975: 64).

Customary rights were defended by symbolic rites. After two deer poachers were arrested in 1721, their fellow poachers banded together "under a mock kingly government," elected "King John" as their leader, invaded Farnham Park, and killed eleven deer. Masked and wearing black gloves, they then "paraded through Farnham in open triumph on market-day" carrying the deer with them. This was only the beginning of the "reign" of King John and his men. They proceeded to the enforcement of popular justice. One gentleman, who had imprisoned a carpenter, was forced to release him; another who, having felled timber, refused the poor their customary right to faggots unless they paid, had the bark stripped off his trees and was warned that unless he returned the money to the poor "he must expect a second visit from King John of the Blacks" (Thompson 1975: 142–4). Here, as in many of the tales of Robin Hood, economic grievances led on to the symbolic contestation of power.

Before the full development of capitalism, there was an intimate connection between social liberty and the popular claim to the freedom of the forest. William Waterson, the vicar of Winkfield from 1717, pointed to this connection antithetically when he wrote: "*Liberty* and *Forest Laws* are incompatible" (Thompson 1975: 49). And we can trace the connection throughout Europe. In 1826, *mardi gras* was celebrated in the small French City of Cholet by public satires which included a mock list of the feudal rights of the "Prince de Ténèbres" and the mock trial and hanging of a vassal for having killed a rabbit (Tilly 1980: 1). As late as 1860, a French judge in Montpellier could complain of a demonstration which combined defiance of the game laws with the celebration of political liberty:

> About twenty people, professional marauders and poachers [these are the judge's labels] got together at the news of the transfer of the communal game warden, the object of their dislike. . . . They went through the streets of the village . . . singing the Marseillaise and other songs of seditious character. . . . They added verses stating a desire for a return to the Republic. (Tilly 1980: 15–16)

It was, indeed, the undermining of forest "liberties" which led Marx to a materialist analysis of history. In the 1859 Preface to *A Critique of Political Economy*, Marx wrote: "In the year 1842–3, as editor of the *Rheinische Zeitung*, I experienced *for the first time* the embarrassment of having to take part in discussions of so-called material interests" (Marx 1977: 388; my italics). These "material interests" were inscribed in the conflict over the feudal rights of Rhineland peasants to gather fallen wood in the forests. As landowners developed new concepts of private property, woodgathering became first a minor offense and then theft, subject to harsh penalties (Carver 1982: 6). Customary rights were pushed aside in the hegemony of the property rights of the bourgeois individual.

After that hegemony, the figure of Robin Hood was transmuted into the sylvan reveler of urban wish-fulfillment. But within a pre- or proto-capitalist mode of production, the tales of Robin Hood condensed the contradictions between "Liberty" and "Forest Laws." In "Robin Hoods progresse to Nottingham" (1620–55), he kills fifteen foresters, and in "Robin Hood and the ranger" (*c.* 1740) forest law becomes an issue of debate. The ranger, as "head-forester," asserts that "these are his majesty's deer," whereas Robin Hood claims that the forest "is my own" where "freely I range." Although the ranger defeats the outlaw in combat, he nevertheless joins Robin Hood's band "singing and dancing . . . in the green wood" (Child 1965: 152–3). In "Robin Hood rescuing three squires," published in various undated eighteenth-century "garlands," the outlaw hears that the squires are to be hanged in Nottingham:

> "O have they parishes burnt?" he said,
> "Or have they ministers slain?
> Or have they robbed any virgin,
> Or with other men's wives have they lain?"
> (Child 1965: 180)

When he learns that they are condemned to die only "for slaying the king's fallow deer," he disguises himself as a palmer or beggar and, after entering Nottingham to offer his services as the hangman, rescues the squires and hangs the sheriff. There are several features of the ballad worth noting. It begins by establishing two different functions of the law: its legitimate punishment of, for example, arsonists or robbers of virgins, and its illegitimate preservation of game for a propertied elite. In defending the rights of poachers, the outlaw assumes the guise of social outcast by dressing in the "ragged and torn" apparel of an "old churl." In other words, he ritually defines himself in terms of the vagabond before he contracts with the sheriff for the post of hangman. Through this contract, the outlaw is inscribed as the official agent of a legal apparatus which, in a carnivalesque reversal, is turned against the sheriff.

Reversals are, of course, structural in the tales of an outlaw who is both *hunter* and *hunted*. As hunter, he challenges the feudal game rights of the aristocracy, killing deer in full view of the ranger or sheriff. Similarly, Breton peasants, a month before the French Revolution, "killed between four and five thousand head of game under the eyes of the warden in the plain of St. Germain" (Canetti 1973: 67), and the Blacks paraded their poached deer through the marketplace of Farnham (Thompson 1975: 142). Here, hunting is conducted as a *public ritual*, deliberately addressed to the gaze of authority. But Robin Hood, like the plebeian rebel, is also hunted as an outlaw, and so he is equated with the very creatures which he hunts. Like them, he must use craft, disguise, and flight. Thus the relation between the misrule of the outlaw and the rule of authority could be rewritten as the relations between animal and human.

At the same time, the opposition of animal to human modeled the opposition of the subordinated to the dominant classes. The peasant carnival "kingdoms" of Romans in 1580, for instance, were named the Hare, the Sheep, and the Capon (a castrated fowl), and their street dances and masquerades "proclaimed that the rich of their town had grown rich at the expense of the poor people" (Le Roy Ladurie 1981: 174). In the revolution of 1525, a German peasant shouted "You harnessed us like oxen to the yoke [and] had my father's hands cut off, because he killed a hare on his own field" to the emperor's daughter, now dressed as a beggar, as she was carried away on a dung cart (Kunzle 1978: 63). This brief episode conceals a complex structure relating human to animal and high to low: the peasants are reduced by the aristocracy to domesticated animals (oxen); a peasant hunter, pursuing an undomesticated animal (a hare), is treated as if *he* was wild and is "tamed" by the cutting off of his hands; finally, in a carnivalesque inversion, an aristocrat is forced to submit to the socially "low" (the status of beggar; the dung cart). Here the subordination of animal to human is inverted through the subordination of authority (the "head") to the lower bodily stratum. But in the iconography of the "World turned upside down," the natural ''hierarchy of humans and animals was itself reversed: deer and hares pursued the hunter; birds ate the sportsman; sheep sheared the shepherd; the ass whipped his laden master; the ox slaughtered the slaughterer'' (Kunzle 1978). And these *adynata* or ''impossible things'' could be reinterpreted as a millennial or revolutionary possibility. One leader in the German peasant revolt of 1525 described himself as ''a beast that usually feeds on roots and wild herbs, but when driven by hunger, sometimes consumes priests, bishops, and fat citizens'' (Kunzle 1978: 63). In the tales of Robin Hood, too, the hunted became the hunter. The outlaw, punished even beyond the margins of the *civitas*, was transformed into the pursuer of sheriffs and bishops. Evading statutory law, he rigorously enforced the liberties of the forest.

The carnivalesque, Robin Hood, and the containment of misrule

In the previous two sections, we have attempted to map out the symbolic system through which the figure of Robin Hood and the practices of carnival intersected with forms of popular protest and action. But in tracing this symbolic system from the late fifteenth century to the middle of the nineteenth century, have we not ignored the fact that the carnivalesque has no fixed *locus* as it is marginalized, excluded, or appropriated in the formation and development of the nation-state? Should we not think of the carnivalesque as a marginal language, subject to the pressures on all marginal languages at a time when competing dialects were being generalized ''into something singular enough to be called *the* vernacular'' (Mullaney 1983: 55)? As the ''core'' regions in Europe (such as Castile, the Ile-de-France, London, and the Home Counties) established linguistic monopolies, so the dominant culture ''discovered'' in the ''anti-languages'' of the peripheries, as Henry VIII ''discovered'' in the Welsh language which he outlawed, ''great Discord Variance Debate Division Murmur and Sedition'' (Hechter 1975: *passim*; Mullaney 1983: 55). Social analysts have attempted to trace a similar process for the carnivalesque, looking for the moment of rupture when it was irredeemably relegated to the marginal, the residual, the ''nostalgic.'' Bakhtin argues that from the beginning of the seventeenth century one can observe ''a narrowing down of the ritual, spectacle, and carnival forms of folk culture, which became small and trivial'' (1968: 33). Yves-Marie Bercé dates the rupture a century later when, under the combined pressure of religious and civic authorities and the disintegration of the rural community, carnival was incorporated into the festivities organized by the aristocracy and urban elites (1976: *passim*). Peter Burke combines the positions of Bakhtin and Bercé, arguing that the ''Triumph of Lent'' depended upon two phases of reform by the elite, the first from 1500 to 1650, the second from 1650 to 1800 (Burke 1978: 207–43). Finally, Alain Faure claims that only in the last decades of the nineteenth century were Parisian festivals taken over and made "respectable" (1978: 167).

What can we learn from these contradictory chronologies? Not, in my view, that change is only in the eye of the beholder. On the contrary, there are conjunctures in which the carnivalesque is radically transformed by the state apparatus, conjunctures whose timing will depend upon the differential development of states. In seventeenth-century England, there was a concerted effort by the monarchy, both before and after the Civil War, to patronize May games and village sports as a means of social control (Stallybrass 198). It was, on the other hand, in the late nineteenth century that the Venezuelan president, Guzmán, supervised "the double transformation of Carnival from a privately-ordered, lower-class event to an officially-ordered, upper-class event; in other words, Carnival as an event of the powerless to Carnival as an event of the powerful" (Lavenda 1980: 469).

There is, though, not *one* irreversible moment when the carnivalesque is finally relegated to the margins of history. (Even in the twentieth-century, we would perhaps be advised to think in terms of displacement, internalization, and dispersal rather than of total appropriation or disintegration.) In searching for the moment when carnival dies, historians rightly avoid the reification of carnival as an immutable psychic or social category. But they do so only at the risk of seeing carnival in terms of a history of permanent decline. This view is, in fact, analogous to the strategy by which the dominant classes rewrite a mode of symbolic contestation as an arcadian dream, acceptable because irretrievably in the past. The historian is, then, in danger of repeating the very categories which most need to be analyzed. These categories are also shared by those literary historians who analyze the "degeneration" of Robin Hood from the "greatness" of the medieval "Gest" to the "cast-iron" ballads of the seventeenth century (see, for example, Nelson 1973: 21–3, 208–10; Child 1965: 42; Dobson and Taylor 1976: 47, 50).

But how should we understand Robin Hood if neither *sub specie aeternitatis* nor *sub specie mortis*? I would argue, following Volosinov, that we need to see Robin Hood as an ideological sign, intersected by "differently orientated social interests" (1973: 23). The meaning of Robin Hood is not given in one originary moment; on the contrary, his meaning is produced and reproduced within the hegemonic process. We have already examined various productions of Robin Hood within popular discourse. I want to turn now to the strategies (partial, contested, and reversible as they were) by which the dominant classes sought to contain the outlaw's misrule. We can divide these strategies into the following categories: opposition or exclusion (the predominant methods of the Church); marginalization or appropriation (the predominant methods of the state).

Strategies of opposition or exclusion

The Church, both before and after the Reformation, used the tales of Robin Hood diacritically to establish the devout in opposition to the wantonly secular. In *Dives and Pauper*, probably written between 1405 and 1410, there is an attack upon those who "gon levir to heryn a tale or a song of robyn hode or of some rubaudry than to heryn messe or matyns" (Dobson and Taylor 1976: 2). In 1508 the priest, Alexander Barclay, translating *The Shyp of Folys*, claimed that he wrote "no jeste ne tale of Robin Hood," and later he declares:

> For goodlie scripture is not worth an hawe,
> But tales are loved ground of ribaudry:
> And many are so blinded with their foly,
> That no scripture thinke they so true nor gode,
> As is a foolish jest of Robin Hode. (Ritson 1887: xxv)

Religious discourse, in other words, used the tales of Robin Hood to construct a semantic field in which ribaldry is opposed to serious texts, folly to wisdom, and jest to devotion. The tales stand as the antithesis of godly order. The antithesis was further elaborated after the Reformation in the attempt to preserve the sacred form from contamination by the pagan and profane. Hugh Latimer, for instance, in a sermon preached before Edward VI in Lent 1549, told of his attempt to preach on "holy day," as he calls it, in a town near London. But, he says, "when I came there, the churche dore was faste locked."

> I tarried there half a houer and more, at last the keye was founde, and one of the parishe commes to me and sayes. Syr thys is a busye daye wyth us, we can not heare you, it is Robyn hoodes daye. The parish are gone a brode to gather for Robyn hoode, I praye you let them not. I was fayne there to geve place to Robyn hoode, I thought my rochet shoulde have bene regarded, thoughe I were not, but it would not serve, it was fayn to geve place to Robyn hoodesmen.
>
> It is no laughynge matter my friends, it is a weepyng matter, a heavy matter, a heavy matter, under the pretence for gatherynge for Robyn hoode, a traytoure, and a thefe, to put out a preacher, to have hys office lesse estemed, to prefer Robyn hod before the ministracion of Gods word. (Latimer 1869: 173–4)

Latimer, of course, is attacking Robin Hood *games*, not tales. But his objections are articulated through semantic oppositions similar to those which characterize treatments of Robin Hood in medieval religious discourse.

In Latimer's view, the "busye daye" of the holiday has displaced the "worcke" of preaching on "holy day," just as the Lincoln green of Robin Hood has displaced the "rochet" or bishop's vestment, the emblem of episcopal authority. More striking still is the inversion of "holy day," for the church door is "fast locked" whilst the "parishe are gone a brode." The sacred itself is excluded in the transformation of holy day to holiday. "Gathering for Robyn hoode" is equated with "put[ting] out a preacher."

Latimer, like most of the godly of the Reformed Church, attempted to reverse this relation between the carnivalesque and the sacred, rewriting the opposition between wilderness and enclosure as a narrative in which textual or festive transgression (ribaldry) led to religious transgression (the avoidance of Church and scripture) which in turn led to social transgression (gatherings outside the surveillance of Church and state). Archbishop Grindal's "Visitation Articles" inquired

> Whether the ministers and church-wardens have suffered any Lords of Misrule, or Summer Lords or Ladies, or any disguised persons, or others, at Christmas or at May-games, or . . . at any other times, to come unreverently into the church or churchyard there to dance or play unseemly parts, with scoffs, jests, wanton gestures, or ribald talk. (Chambers 1903: 383)

The triumph of Lent required the elimination of carnival.

Strategies of marginalization or appropriation

While the church attempted to exclude Robin Hood, the state sought both to limit the *scope* of the outlaw's misrule and to appropriate misrule as the license which the monarch himself both enjoyed and had the power to grant. As we shall see, both Henry VIII and Charles II acted out the role of outlaw. But during the seventeenth century, we can also observe a strategy by which Robin Hood was marginalized, reduced to a renegade in need of communal correction.

In *The Pinder of Wakefield*, for instance, Robin Hood is transformed into the sidekick of George, the pinner. The chapbook, published in 1632, describes the outlaw's band as trespassing upon the fields of Wakefield, "tearing downe of hedges, making new pathes over the Corne, cutting downe of stiles, carrying long staves on their shoulders, breaking all the good orders that George had made" (Horsman 1956: 67). It is the task of the pinner to impound stray animals which trample down the crops. Thus Robin Hood and his men are structurally equivalent to stray animals, destroying the "good orders" both practiced in and symbolized by the enclosure. The pinner triumphs over the outlaw, here as in other versions of their meeting. In *Strappado for the Divell* (1615), for instance, Richard Brathwaite writes of

> The pindar's valour, and how firme he stood
> In the townes defence, gainst th' rebel Robin Hood,

How stoutly he behaved himselfe, and would
In spite of Robin, bring his horse to th' fold.
(Ritson 1887: xxvii)

George, like Robin Hood, leads a band whose main pursuits are eating, drinking, fighting, and other ''pastimes and merriments'' (Horsman 1956: 9). But unlike the ''rebel'' outlaw, George celebrates the carnivalesque only within the limits provided by ''good orders.'' Carnival is reduced to the policing of domestic relations. To ridicule a ''scold,'' George initiates a charivari; his followers dress a boy ''in womens apparell like the woman, and a man like her husband, and put them both on a horse. . . . Thorow the Towne thus they rid, the woman beating the man, and scolding him terribly'' (Horseman 1956: 13). To George is also attributed the invention of the ducking stool to punish scolds. When he is not victimizing the insubordinate, he is defending the elite, preventing thieves from robbing a gentleman's house. The carnivalesque is used to reaffirm the social order. Again, in *The Comedy of George a Green*, the pinner is the staunch upholder of ''good orders,'' striking the rebel Earl of Kendal, and forcing Kendal's follower, Mannering, to eat his own commission. Not surprisingly, Robin Hood is introduced into the play only as a device to demonstrate George's strength and as a loyal subject of the king.

We may note, though, that the reinscription of Robin Hood within the ideological enclosure of the nation-state is not the only possible outcome of his confrontation with the pinner. In the ballad ''The Jolly Pinder of Wakefield'' (*c.* 1650–60), Robin Hood again meets his match in the pinner. Here, though, the combat is the prelude to the incorporation of the pinner into the outlaw's band. More important, the pinner's incorporation necessitates a reconsideration of his status. For at the beginning of the ballad, he seems to be an independent yeoman. He boasts that he is a match not only for knight and squire but even for any ''baron that is so bold,/Dare make a trespasse to the town of Wakefield'' (Dobson and Taylor 1976: 147). In the final stanza, though, the pinner promises to defect from his ''craft,'' complaining of his servitude:

If Michaelmas day were once come and gone
And my master had paid me my fee,
Then would I set as little by him
As my master doth set by me.

The ballad constructs two different versions of the pinner. The first is related to the pinner of *The Comedy*, who owns allegiance to no one but the monarch. But the second will ''plod to the green wood'' to escape from his master. The ''pinfold'' of which the pinner is the proud keeper is displaced by the pinfold of service in which he is held in bondage. He ends, like Robin Hood, by choosing the life of a thief or rebel.

At stake in the different versions of the Pinner of Wakefield is the concept of enclosure. In the ballad, the enclosure of space, like the enclosing of time in the contracting of the pinner's labor, is ruptured by the utopian space and time of the forest. In the town, the pinner is dependent upon a "covenant" with a fixed term (Michaelmas); but as Orlando says in *As You Like It*, "there's no clock in the forest" (Shakespeare 1974: 386). The occupation of the pinner in the play and the chapbook, on the other hand, becomes a model of "good orders," producing a series of homologies between the fold, the hierarchized neighborliness of the local community, and the "enclosure" of the nation-state.

Martin Parker similarly attempts to incorporate Robin Hood into the regulated time of the king's rule in "A true tale of Robin Hood" (1632). But Parker's simultaneous loyalty to the crown and admiration of the outlaw result in textual contradictions. At one point the outlaw robs the king's tax-gatherers, "Not dreading law"; at another, he is said to attack only "such as broke the lawes" and

> He wished well unto the king,
>   And prayed still for his health,
> And never practised any thing
>   Against the common wealth
>   (Child 1965:229–30)

But Parker's main method of "correcting" the outlaw is to set him firmly in the past:

> We that live in these latter dayes
>   Of civill government,
> If neede be, have a hundred wayes
>   Such outlawes to prevent.
>
> In those dayes men more barbarous were,
>   And lived lesse in awe:
> Now, God be thanked! people feare
>   More to offend the law.

Parker's desire to historicize Robin Hood is made clear in the title of the ballad:

> A True Tale of Robin Hood, or a briefe touch of the life and death of that Renowned Outlaw, Robert Earl of Huntington vulgarly called Robbin Hood, who lived and died in A.D. 1198, being the 9. yeare of the reign of King Richard the first. . . . Carefully collected out of the truest Writers of our English Chronicles. And published for the satisfaction of those who desire to see Truth purged from falsehood. (Child 1865: 227)

What are the marks of Parker's "Truth"? The elevation of Robin Hood to the peerage (separating him off from the "vulgar") in imitation of Anthony Munday's plays on the Earl of Huntington (Munday 1964, 1965); the antiquarian and patriotic emphasis on English chronicles; a scrupulous attention to dating. The legendary outlaw (of whose actual existence there is no reliable evidence) is rewritten in the scholarly historical mode of emergent nationalism with its enthusiastic recording of the barbarous but heroic deeds of a past which can be admired because located at a safe distance in time.

The outlaw could also be tamed by confining him to the licensed misrule of official feast dates, times which could be patronized and presided over by the new aristocracy. Sir John Paston, for instance, employed a servant "to pleye Seynt Jorge and Robyn Hood and the Shryff of Nottyngham" (Dobson and Taylor 1976: 38). In 1559, Elizabeth I patronized 'a May-game . . . with a gyant, and drumes and gunes . . . and then sant Gorge and the dragon, the mores dansse, and after Robyn Hode and lytyll John, and M[aid Marion] and frere Tuke." On June 23, the May game was performed "a-bout London" but the following day it was presented "at Grenwyche, playing a-for the Quen and the consell" (Machyn 1848: 201). The monarchy also appropriated to itself more directly the "benign" misrule of the outlaw. In 1510, Henry VIII and eleven nobles broke into the Queen's chamber dressed "like outlawes, or Robyn hodes men," transforming the "gests" of the forest into the theater of patrician sexual licence. In May 1515, the king was met at Shooter's Hill by "Robyn Hood" and his men "clothed all in grene," and he went with them "into the grene wood, and to se how the outlawes lyve" (Hall 1809: 520, 582). Charles II was even more closely connected to the outlaw and the carnivalesque. In "Robin Whood revived, a cavalier song," Charles is identified as Robin Hood (Dobson and Taylor 1976: 238) and after the Restoration he was frequently portrayed as the May king, his effigy sometimes being used in the May games (Marcus 1979: 53). He was proclaimed king on May 8, one of St George's festival days, and he timed his triumphal entry into London to coincide with his birthday, May 29.

One of the more extraordinary documents of the Restoration is a short comedy entitled *Robin Hood and his Crew of Souldiers*, which was acted at Nottingham "on the day of His sacred Majesties Coronation," April 23 1661 (Dobson and Taylor 1976: 237–42). The play begins with a royal messenger demanding of the outlaws "a chearfull and ready submission to his Majesties Laws," with a promise of future obedience. The outlaws are suitably shocked. William Scadlocke declares:

> If this geare takes them we may turn our Bows into Fiddle-sticks, or strangle ourselves in the strings, for the daies of warre and wantonness will be done. . . . Wakes and Bear-baitings, and a little Cudgel-play must be all our comfort. . . . We must not dream of Venison, but be content like the King's liege-people with crusts and mouldy Cheese.

Whilst Scadlocke emphasizes extra-legal pleasures, Little John contests the notion of statutory itself: "bounteous nature ties not her selfe to rules of State, or the hard Laws that cruell men impose." And Robin scorns the law-abiding who "have grown inamoured of their Chaines, and caressed their slavery, and doat upon their hateful Bondage." The royal messenger concludes this disquisition upon the foundations of law by claiming that legal sanctions insure a form of freedom: "That by the Laws which careful Princes make, we are commanded to do well and live virtuously, free both from giving and receiving injuries, is not to be esteemed slav'ry but priviledge." The debate is concluded, though, not by the triumph of the messenger's arguments but by the formal structure of the play in terms of anti-masque and masque. Political conflict is resolved, as in the masque where, by a dramatic shift of scenes, the wild (anti-masque) is simply displaced by the civilized, through a "sudden change" in which Robin's aesthetic perceptions are reformed. He immediately becomes "quite another man," "dwarfd" and shrunk to a "low pitch" by "some power great and uncommon." The play ends with the submissive outlaws singing "hearty Wishes, health unto our King" (Dobson and Taylor 1976: 241).

What were the functions of incorporating the outlaw as loyal subject? First, it colonized a threat by reducing misrule to the chaotic antithesis of the "privilege" of rule. Second, it naturalized misrule by making it a seasonal and "mythic" moment which could no more challenge the inevitability of rule than the shroving of carnival could prevent the shriving of Lent. Third, it presented the monarchy as a source of *license* rather than of control. It was with this in mind that the Duke of Newcastle wrote to Charles II, encouraging the revival of the maypole, the morris dance, and "all the olde Holyedayes with their Mirth" (Strong 1903: 226–7). In this way, political and social hierarchy could be effaced by the elision of monarchy, play, and carnival in the ideological construction of England as a band of brothers.

Conclusion

We may now attempt to represent schematically how the figure of Robin Hood was constituted, reproduced, and contested within the hegemonic process. I borrow the diagram presented in Figure 2.1 from Pierre Bourdieu (1977: 168), although I shall appropriate it for the purposes of my own argument.

1. *Heterodoxy is produced by orthodoxy.* Orthodoxy produces Robin Hood as a form of licensed misrule, colonizing conflict through ritual contestation organized from above and separated off in time and space from the quotidian. It may also appropriate and psychologize Robin Hood for itself, as when Henry VIII and Charles II transformed the outlaw's (social) "liberty" into a sign of their own psychic (libidinal) power.

2. *Heterodoxy is excluded by orthodoxy*. Orthodoxy suppresses Robin Hood and his misrule. This was the strategy most frequently pursued by the "godly sort," but it was subject to two major weaknesses: it could not be enforced without adequate central or local policing; even when enforceable, it tended to relegate misrule to a space outside official surveillance.
3. *Heterodoxy negotiates/contests orthodoxy*. At its weakest, heterodoxy contests only the *workings* of orthodoxy, not its principles. This perspective constructs a Robin Hood who punishes oppressive landlords, but is loyal to the monarch, whose liberality and hospitality are the virtues of the "good" squire, and who "seeks to establish or to re-establish justice or 'the old ways', that is to say, fair dealing in a society of oppression" (Hobsbawm 1981: 55). In a pre-capitalist society, there is also a tendency for the outlaw celebrated by the subordinated classes to embody "the only familiar model of 'freedom', i.e., the status of nobility" (Hobsbawm 1981: 37n). Even at its weakest, though, heterodoxy challenges the *absolute* authority of the dominant and contests the development of new forms of exploitation, if still appealing to paternalism (Thompson 1971: 83–8).

   Moreover, what looks like "conservative" negotiation can often be reinterpreted as: the necessary rhetorical disguise of a group threatened by censorship or reprisals; the tactical construction of "the old ways" as a means of criticizing existing domination; total submission to "the king," when he is primarily a symbolic representation of popular justice to legitimate insurrection against the everyday workings of economic, social, and political oppression.

   We are accustomed to the way in which a class fraction of the dominate classes could use popular protest as a method of pursuing its own intra-class rivalries. But some of the Robin Hood ballads *reverse* this strategy: for instance, the outlaw seeks favor with the king to pardon an attack upon the king's tax gatherers or the sheriff, or he wins the queen's support against a bishop or even against the king (displacing the act of cuckolding from the sexual to the social domain?) (Child 1965: 220–2, 196–204, 205–7). In other words, heterodoxy does not necessarily contest or negotiate with a *unified* orthodoxy. On the contrary, it can exploit the fissures within the dominant, thus exposing "the universe of unquestioned assumptions" concealed by the naturalizations of orthodoxy.
4. *Heterodoxy interrogates the boundaries between Doxa and Opinion*. It is difficult to separate this category from the previous one in any concrete analysis of the hegemonic process in early modern England. Did challenges to the game laws work within a reformist problematic, appealing against specific grievances, or did they subvert the very concept of property through which class differentiation was produced? No doubt,

contestation of the notion of property often evaporated with the paternalist amelioration of more immediate hardships; but, vice versa, local grievances could develop into a counter-ideology which challenged both the existing legal system and the monopoly of the propertied in the making of laws. But perhaps we should express this in a way which does not denigrate the "merely" local. For through the figure of Robin Hood, the practical politics of village and urban life could be projected as, and legitimated by, the utopian space of the outlaw; and, vice versa, the liberties of the forest could become the standard by which the "unquestioned assumptions" of the quotidian were judged and contested.

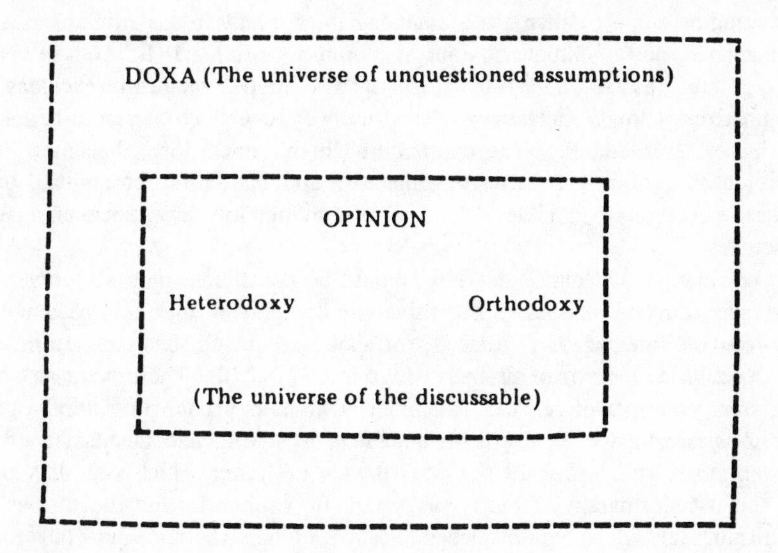

Figure 2.1: From this simple but polysemic diagram we can develop the following models of the relation between orthodoxy and heterodoxy, inlaws and outlaws:

Strategies of the dominant:
1. Heterodoxy is *produced* by orthodoxy
2. Heterodoxy is *excluded* by orthodoxy

Strategies of the subordinated:
1. Heterodoxy *negotiates/contests* orthodoxy
2. Heterodoxy *interrogates the boundaries* between Doxa and Opinion, thus questioning the terrain on which orthodoxy and heterodoxy are alike constructed.

We claimed above that Robin Hood is an ideological sign where "differently orientated social interests" intersect (Volosinov 1978: 23). In early modern England, this intersection of interests was nowhere more important than in the concepts of "play" and "law." The dominant classes could appropriate Robin Hood to the extent that they could reinscribe his lawlessness as play. Bishop Fuller, for instance, in his *History of the Worthies of England* (1662) vacillates in his treatment of the outlaw. From the point of view of "law" (or the legal apparatus of the state), Robin Hood deserves "the gallows for his many robberies": "[W]ho made him a judge? Or gave him a commission to take where it might best be spared, and give where it was most wanted?" But then, Fuller decides, Robin Hood was after all "rather a merry than a mischievous thief" and so his offenses can be categorized not as "thievery" but as "pranks" (Fuller 1840: 568–9). *Mere* play: thus, as Allon White has shown, "high" language excludes or stigmatizes "low" languages "as 'funny', 'one-sided', 'not-to-be-taken-seriously'" (1983: 14). The conflict of "high" and "low" becomes ritual play, a ruse of power through which the dominant can "consolidate itself more effectively" (White 1982: 60), leaving the law uncontaminated, uncontested.

The relation of "play" to "law" could be constituted quite differently by the subordinated classes. In carnivalesque "play," the law became the subject of interrogation. Indeed, official law could be seen from this perspective as a form of *disorder* (Ingram 1977: 110). There were also two different concepts of justice: "an older, Christian version of natural rights, which guaranteed even the poorest men at least life; and the justice of the law of property" (Hay 1975a: 35). It was the latter which was developed in the transformation of the legal system by capitalism and the former, as a result, tended to be displaced from religious rhetoric into the social practices of the charivari and food and enclosure riots. Open conflict between "natural rights" and "the law of property" became the norm. Plebeian culture regarded the game laws, for example, as "illegal." One magistrate recorded that the rural population "say God has made the game of the land free and left it free" (Munsche 1981: 96); another noted, "it is the common defence of a poacher, that it is very hard that he should be punished for taking what he had as good a right to as any other man" (Hay 1975b: 207). In the ballads of Robin Hood, as Roberta Kevelson puts it, the question of "Who wins the game?" becomes an argument over "Who owns the game?" (1977: 67). The "play" of carnival in the "jests" of the outlaw is, at the same time, a disquisition on game laws and property. As we have argued, the ballads' *rhetoric* of violence is structured by, even as it helps to structure, a rhetoric of *violence* (the symbolic rituals of poaching and the destruction of enclosures).

At the same time, plebeian "play" was both a means of unmasking legal violence (the violence of which the state claimed a monopoly) and of producing alternative definitions of "law." It remained for the era of capitalist domination to separate "law" and "play" into totally discrete categories. Only then would symbolic action ("leisure") be appointed its own domain and its own specialists, "liberated" from "real" action (economic, political, and legal practices). But for those who had a "dream of Venison" instead of the "crusts and mouldy cheese" of the "King's liege-people," for those who danced "in the myre for veri pure joye," for those who were "drunk with the cup of liberty," "Robin hodes daye" was, indeed, "a busye daye."

References

Agulhon, Maurice (1981) *Marianne into Battle: Republican Imagery and Symbolism in France, 1798–1880*. Trans. Janet Lloyd. Cambridge: Cambridge University Press.

Arnold, Matthew (1932) *Culture and Anarchy*. Ed. J. Dover Wilson. Cambridge: Cambridge University Press.

Bakhtin, Mikhail (1968) *Rabelais and His World*. Trans. Hélène Iswolsky. Cambridge, Mass.: MIT Press.

Beer, Barrett L. (ed.) (1976) "'The Commoyson in Norfolk 1549': A Narrative of Popular Rebellion in Sixteenth-Century England." *Journal of Medieval and Renaissance Studies 6(1)*: 73–99.

——— (1982) *Rebellion and Riot: Popular Disorder in England during the Reign of Edward VI*. Kent, Ohio: Kent State University Press.

Bercé, Yves-Marie (1976) *Fête et révolte: Des mentalités populaires du XVIe au XVIIIe siècle*. Paris: Hachette.

Bonnain-Moerdyk, R. and Moerdyk, D. (1977) "A propos du charivari: discours bourgeois et coutumes populaires," *Annales: Economies, Sociétés, Civilisations* 32(2): 381–98.

Bourdieu, Pierre (1977) *Outline of a Theory of Practice*. Trans. Richard Nice. Cambridge: Cambridge University Press.

Burguière, André (1980) "The Charivari and Religious Repression in France during the Ancien Regime," in Robert Wheaton and Tamara K. Hareven (eds) *Family and Sexuality in French History*. Philadelphia: University of Pennsylvania Press: 84–110.

Burke, Peter (1978) *Popular Culture in Early Modern Europe*. London: Temple Smith.

Canetti, Elias (1973) *Crowds and Power*. Trans. C. Stewart. Harmondsworth: Penguin.

Carver, Terrell (1982) *Marx's Social Theory*. Oxford: Oxford University Press.

Chambers, E.K. (1903) *The Medieval Stage*, 1, Oxford: Clarendon.

Child, Francis James (1965) *The English and Scottish Popular Ballads*, 3. New York: Dover.

Davis, Natalie Zemon (1978) "Women On Top: Symbolic Sexual Inversion and Political Disorder in Early Modern Europe," in Barbara B. Babcock (ed.) *The Reversible World*. Ithaca, NY: Cornell University Press: 147–90.

Dobson, R.B. and Taylor, J. (1976) *Rymes of Robyn Hood*. Pittsburgh: University of Pittsburgh Press.

Douglas, Mary (1978) *Purity and Danger: An Analysis of Concepts of Pollution and Taboo*. London: Routledge & Kegan Paul.

Faure, Alain (1978) *Paris Carême-prenant: du Carnaval à Paris au XIXe siècle*. Paris: Hachette.
Fletcher, Anthony (1973) *Tudor Rebellions*. London: Longman.
Fuller, Thomas (1840) *The History of the Worthies of England*. London: Thomas Tegg.
Hall, Edward (1809) *Hall's Chronicle*. London: J. Johnson.
Hay, Douglas (1975a) "Property, Authority and the Criminal Law," in Douglas Hay *et al.* (eds) *Albion's Fatal Tree: Crime and Society in Eighteenth-Century England*. New York: Pantheon: 17–63.
——— (1975b) "Poaching and the Game Laws on Cannock Chase," in Douglas Hay *et al.* (eds) *Albion's Fatal Tree: Crime and Society in Eighteenth-Century England*. New York: Pantheon: 189–253.
Hechter, Michael (1975) *Internal Colonialism: The Celtic Fringe in British National Development, 1536–1966*. London: Routledge & Kegan Paul.
Hill, Christopher (1975) *The World Turned Upside Down: Radical Ideas During the English Revolution*. Harmondsworth: Penguin.
Hobsbawm, Eric (1981) *Bandits*. New York: Pantheon.
Holinshed, Raphael (1808) *Holinshed's Chronicles of England, Scotland and Ireland*, 3. London: J. Johnson.
Horsman, E.A. (1956) (ed.) *The Pinder of Wakefield*. Liverpool: Liverpool University Press.
Ingram, M.J. (1977) "Communities and Courts: Law and Disorder in Early-Seventeenth-Century Wiltshire," in J.S. Cockburn (ed.) *Crime in England*. Princeton: Princeton University Press: 110–34.
Kevelson, Roberta (1977) *Inlaws/Outlaws. A Semiotics of Systemic Interaction: "Robin Hood and the King's Law."* Studies in Semiotics 9. Bloomington: Indiana University Press.
Kirby, Chester and Ethyn (1931) "The Stuart Game Prerogative." *English Historical Review* 46: 239–54.
Kristeva, Julia (1980) *Desire in Language: A Semiotic Approach to Literature and Art*. Trans. Thomas Gora, Alice Jardine, and Leon S. Roudiez. New York: Columbia University Press.
Kunzle, David (1978) "World Upside Down: The Iconography of a European Broadsheet Type," in Barbara B. Babcock (ed.) *The Reversible World*. Ithaca, NY: Cornell University Press: 39–94.
Latimer, Hugh (1895) *Seven Sermons before Edward VI*. Ed. Edward Arber. London: Constable.
Lavenda, Robert H. (1980) "The Festival of Progress: The Globalizing World-System and the Transformation of the Caracas Carnival." *Journal of Popular Culture* 13(3): 465–75.
Le Roy Ladurie, Emmanuel (1981) *Carnival in Romans: A People's Uprising at Romans 1579–1580*. Trans. Mary Feeney. Harmondsworth: Penguin.
Machyn, Henry (1848) *The Diary of Henry Machyn*. Ed. John Gough Nichols. London: J.B. Nichols.
Marcus, Leah S. (1979) "Herrick's *Hesperides* and the 'Proclamation made for May.'" *Studies in Philology* 76: 49–74.
Marvell, Andrew (1971) *The Poems and Letters of Andrew Marvell*. Ed. H.M. Margoliouth, revised Pierre Legouis and E.E. Duncan-Jones. Oxford: Clarendon.
Marx, Karl (1977) *Karl Marx: Selected Writings*. Ed. David McLellan. Oxford: Oxford University Press.
Mellinkoff, Ruth (1973) "Riding Backwards: Theme of Humiliation and Symbol of Evil." *Viator* 4: 153–76.

Mill, Anna Jean (1927) *Medieval Plays in Scotland*. Edinburgh: Blackwood.
Mullaney, Steven (1983) "Strange Things, Gross Terms, Curious Customs: The Rehearsal of Cultures in the Late Renaissance." *Representations* 3: 40–67.
Munday, Anthony (1964) *The Downfall of Robert, Earl of Huntingdon*. Ed. John C. Meagher. Oxford: Malone Society.
——— (1965) *The Death of Robert, Earl of Huntingdon*. Ed. John C. Meagher. Oxford: Malone Society.
Munsche, P.B. (1981) "The Gamekeeper and English Rural Society, 1660–1830." *Journal of British Studies* 20(2): 82–105.
Nelson, Malcolm A. (1973) *The Robin Hood Tradition in the English Renaissance*. Salzburg Studies in English Literature, 14. Salzburg: Institut für Englische Sprache und Literatur.
Phythian-Adams, Charles (1979) *Desolation of a City: Coventry and the Urban Crisis of the Late Middle Ages*. Cambridge: Cambridge University Press.
Ritson, Joseph (1887) *Robin Hood: A Collection of All the Ancient Poems*, 1. London: John C. Nimmo.
Scribner, Bob (1978) "Reformation, Carnival, and the World Turned Upside-Down." *Social History* 3(3): 303–29.
Shakespeare, William (1974) *The Riverside Shakespeare*. Ed. G. Blakemore Evans. Boston: Houghton Mifflin.
Sharp, Buchanan (1980) *In Contempt of All Authority: Rural Artisans and Riot in the West of England, 1586–1660*. Berkeley: University of California Press.
Simeone, W.E. (1951) "The May Games and the Robin Hood Legend." *Journal of American Folklore* 64: 265–74.
Spenser, Edmund (1912) *Poetical Works*. Ed. J.C. Smith and E. De Selincourt. London: Oxford University Press.
Stallybrass, Peter (1982) "Carnival Contained: Patrician Festivity and Nationalism in Early Modern England." Unpublished ms.
Stow, John (1980) *A Survey of London*, 1. Ed. Charles Lethbridge Kingsford. Oxford: Clarendon.
Strong, S.A. (1903) *A Catalogue of Letters at Welbeck*. London: John Murray.
Stubbes, Philip (1877) *The Anatomie of Abuses*. Ed. F.J. Furnivall. London: New Shakespeare Society.
Thirsk, Joan and Cooper, J.P. (1972) *Seventeenth-Century Economic Documents*. Oxford: Clarendon.
Thomas, Keith (ed.) (1966) "Another Digger Broadside." *Past and Present* 42: 57–68.
——— (1976) *Rule and Misrule in the Schools of Early Modern England*. Reading: Reading University Press.
Thompson, E.P. (1963) *The Making of the English Working Class*. London: Gollancz.
——— (1971) "The Moral Economy of the English Crowd in the Eighteenth Century." *Past and Present* 50: 76–136.
——— (1975) *Whigs and Hunters: The Origin of the Black Act*. New York: Pantheon.
Tilly, Charles (1980) *Charivaris, Repertoires, and Politics*. Center for Research on Social Organization Working Paper no. 214. Ann Arbor: University of Michigan.
Volosinov, V.N. (1973) *Marxism and the Philosophy of Language*. Trans. Ladislav Matejka and I.R. Titunik. New York: Seminar Press.
Walter, John (1980) "Grain Riots and Popular Attitudes to the Law: Maldon and the Crisis of 1629," in John Brewer and John Styles (eds) *An Ungovernable People: The English and their Law in the Seventeenth and Eighteenth Centuries*. New Brunswick: Rutgers University Press.

White, Allon (1982) ''Pigs and Pierrots: The Politics of Transgression in Modern Fiction.'' *Raritan* 2(2): 51–70.

——— (1983) '''The Dismal Sacred Word'. Academic Language and the Social Reproduction of Seriousness.'' *Literature, Teaching, Politics* 2: 4–15.

Wrightson, Keith (1982) *English Society 1580–1680*. London: Hutchinson.

3

Violence done to women on the Renaissance stage

Leonard Tennenhouse

I

The following essay deals with the particular form of violence directed against the aristocratic female body in Jacobean drama. I will be considering that body as well as its treatment as a discursive practice. I do not take it to be either a "real" body, or a "mere" representation of the female, but rather an actor playing the part of an aristocratic woman. That such a practice existed there can be no doubt. Around the year 1604, dramatists of all sorts suddenly felt it appropriate to torture and murder aristocratic female characters in a shocking and ritualistic manner. This assault was quite unlike anything seen on the Elizabethan stage – even at its most Senecan. If any statement holds true about violence done to these female characters, it is that such violence is never simply violence done to them as women. It is always violence done to one occupying a particular position in the social body as it was conceived at the time. Given my reliance on these terms, then, let me briefly explain what I mean by "the social body" and "the aristocratic body."

The "body" is a problematic term to begin with, because in recent years it has been so widely used with varying degrees of looseness and precision. Indeed, we have reason to think the body itself has been used in most cultures and at different times as a figure with which to think out the relationship between individual bodies and that aggregate of bodies called society. Discussing the anthropology of the body, Jean Comaroff summarizes the current view on the construction of the body as a social subject: "Through socialization, the 'person' is constituted in the social image, tuned, in practice, to the coherent system of meanings that lies silently within the objectives of a given world" (1985: 7). Further, that body is always inscribed with and used to think about social relationships. Even in modern cultures, the body is not so much a natural object as an image or sign we use to understand ourselves as selves. We carry a body around in our heads that governs the ways in which we represent ourselves to the world. We also carry about a social body image which we use to represent the world to ourselves. For these two images are actually a single cultural formation.[1]

In truth, I cannot decide which is the more difficult project – to understand our own bodies as cultural objects or to determine what members of another culture – in this case, one several centuries earlier than ours – imagined their bodies to be. Such is the nature of cultural counter-transference where bodies are concerned that we cannot know another's without having some idea of the conceptual contours and ideological projects inscribed on our own. About the middle-class body, Nancy Armstrong has written:

> Ours is a social body divided in two along lines of gender: a male body corresponding to the masculine domain of productive labor and a female body corresponding to the feminine domain of the household. While all cultures both make things and reproduce people, industrial societies are unique in their way of gendering labor, the space in which it occurs, and the bodies performing it. (1988: 4).

This is not to say that the difference between male and female was insignificant to the people for whom Shakespeare's company performed. It obviously mattered. But more important still than the question of whether one were a male or a female was the question of membership of the aristocracy. When Bakhtin sought to describe the concept of the body one needed in order to make sense of Rabelais, he employed a primary basis for difference other than gender. He called it the "mass body" and set that body in opposition to the classical or elite body. Where the classical body, or what I shall call the aristocratic body, was ordered, hierarchized, impermeable, and pure, the mass body was open and protruding, riotous, heterogeneous, sensual, and renewable.

Obviously gender plays a significant role in such a social body, but it does not determine the distinction of first importance in maintaining an aristocratic culture. In a culture that understands difference first and foremost in terms of whether one belongs inside or outside a privileged community, gender is one more way of marking difference between the elite and mass bodies. One way of understanding the difference between the symbolic properties of Renaissance women and those inhering in a culture where gender overrides class in determining identity is to see how this difference was inscribed on the body. Anatomy books, midwifery guides, and manuals on obstetrics represented male and female as possessing essentially the same body (Eccles 1976; Martin 1987: 27–32; Laqueur 1986: 4–16). The two were structurally homologous, and difference was understood to rest on the degree of heat one possessed, which varied according to whether the sexual organs were inside or outside of the body. Thus one cannot speak about gender in the Renaissance without first speaking about political hierarchy. The aristocratic female was automatically superior to a man from the lower ranks. Within the aristocratic body, however, she occupied a position of lesser degree in relation to the male of the same station.

But while this is a good rule of thumb, it does not tell us all that much about the symbolic properties of Shakespeare's women or the various fates that befell them on the stage. Never a stable entity, the aristocratic body was constantly changing. Not only did its size and membership change over time, but individual monarchs also left their respective marks on representations of that body. Under Elizabeth, the highest position, that of the patron of patrons, was occupied by a woman, and so we may speculate it made perfect sense to represent the aristocratic body as a female body. Indeed, as I will explain, Elizabeth insisted upon it. Under James, however, this gender theme was revised and incorporated in a new image of the body politic. On the one hand, we find romances and tragi-comedies that celebrate the reunion of an originary family under a chastened monarch/father. No matter what human forces seek to dismember this body, a miraculous force watches over members of the royal body and ensures their mutual attraction. Tragedies, I will argue, approached the same problem of revision from an entirely different angle. They stripped away the very qualities that had distinguished heroines of just a few years before. Thus on the Jacobean stage we see aristocratic women punished for possessing the very features that empowered such characters in Elizabethan romantic comedy. The ritual purification of these bodies did not simply give vent to misogynistic impulses (although I am sure it did that as well, and indeed continues to do so); it also revised the political iconography identified with an earlier monarchy, which was understood by English monarchs to be a very real instrument of their power. Only this, I believe, could have made the assault on the Elizabethan style of female so pervasive.

It is difficult to think of a Renaissance tragedy in which at least one woman is not threatened with mutilation, rape, or murder. Her torture and death provide the explicit and exquisite dénouement and centerpiece of the play in question. Yet despite concerted efforts to historicize the literary past, criticism has done little to account either for the pervasiveness of such violence or for the gender of its victims. That the body of an aristocratic female was the centerpiece of the spectacles of violence on stage had everything, in the Elizabethan period, to do with the Queen herself. She constantly encouraged an equation to be made between the health of her body, its wellbeing and integrity, and that of the state. During her reign, this iconic identification between the queen's body and the land was such that the violence done to one was the same as violence to the other. Thus the theater regularly staged scenes of violence and disorder in order to materialize an opposition to the monarch over which monarchy asserted its order. On the Jacobean stage, however, the aristocratic female having acquired this usage had to be both different from the king's body and yet essential to the purity of the aristocratic community. Once again, she was the site on which to stage an assault on the monarch. As a source of pollution, she empowered the monarch by subjugating her in a ritual that purified the community.

Hamlet shall be a test case for this proposition, because in that play there are two forms of violence, both indicative of earlier cultural practices, and each centering on a female. In the assault on Gertrude, we find a characteristic Elizabethan representation of violence which threatens to dismember the state by internal division. By way of contrast, the Player Queen episode imagines a different model of violence, one more characteristic of the assault against women found on the Jacobean stage. Before turning to *Hamlet*, it is first necessary to suggest how the body of the aristocratic female on stage was used to produce such a political literacy.

II

Elizabeth Tudor knew the power of display. She also knew how to display her power as queen. But this is not to say that even so powerful a monarch as she could determine the conditions for effectively displaying political power. Upon her accession, if not well before, Elizabeth found herself thoroughly inscribed within a system of political meaning. Marie Axton explains:

> for the purposes of the law it was found necessary by 1561 to endow the Queen with two bodies: a *body natural* and a *body politic*. (This body politic should not be confused with the old metaphor of the realm as a great body composed of many with the king as a head. The ideas are related but distinct.) The body politic was supposed to be *contained within the natural body of the Queen*. When lawyers spoke of this body politic they referred to a specific quality: the essence of a *corporate perpetuity*. The Queen's natural body was subject to infancy, error, and old age; her body politic . . . was held to be unerring and immortal. (1977: 12; my italics)

The "lawyers," Axton observes, "were unable or unwilling to separate state and monarch" (1977: 12). Elizabeth insisted upon representing her body as one and the same as England. She made this equation on the grounds that her natural body both contained and stood for the mystical power of blood which had traditionally governed the land and made it English. The concept of the body politic was redefined in certain characteristically Elizabethan ways as it became that of a female patriarch. According to the English form of primogeniture, a female could legitimately and fully embody the power of the patriarch. That power was in her and nowhere else so long as she sat on the throne. In being patriarchal, state power was not necessarily male, for Elizabeth was represented and treated as a female. Thus we may conclude that Elizabethans saw the state as no less patriarchal for being embodied as a female, and they saw the queen as no less female for possessing patriarchal powers. In other words, the idea of a female patriarch appears to have posed no contradiction in terms of Elizabethan culture.

The queen's body was displayed in official portraits, on coins, in the royal coat of arms placed in all the churches of England, and in her official passages through London and royal progresses in the countryside (Strong 1963; Phillips 1973: 119). In the context of her considerable entourage, Elizabeth's very presence called forth elaborate pageants, tributes, opulent shows of all kinds, speeches, orations, and the presentation of gifts, these to be witnessed by large numbers of people.[2] Of particular importance was the role that the public theater played in displaying and idealizing forms of power that were grounded in the value and importance of the aristocratic female body. It is in this sense, I shall argue, that Shakespeare's drama played an active role in the political life of Renaissance England. In arguing for its historical significance, then, I do not want to privilege a topical meaning as paramount in understanding the success of Elizabethan drama. Rather, I want to suggest that the drama for which Elizabethan culture is known offered one of the more important means of controlling how various people imagined state power and understood themselves in relation to it. Given the importance of displaying the aristocratic female body – the most powerful manifestation of which was the appearance of the queen herself – the theater was never more political than when it called attention to the body of an aristocratic female. Elizabeth and her people understood the display of her body in terms of those practices which identified the monarch's body with English power in all its guises.

But a strategy that enabled this unprecedented consolidation of English power in the monarch and her court necessarily gave rise to a major political crisis by the late 1590s with the obvious decrepitude of Elizabeth's body. Visible signs that her natural body was failing called into question the relationship between the queen's two bodies upon which hinged in turn the monarch's symbolic control over England. The crisis brought on by the loss of symbolic power that would accompany her aging and death appears to have been resolved by a shift towards representations that placed compensatory stress on the monarch's body metaphysical. To foreground the continuity of patriarchal power, writers and performances of all kinds emphasized the masculine nature of legitimate political power and, at the same time, began to imagine the aristocratic female body as having the potential to disrupt the flow of power from one male to another. It was in the public theater and the Inns of Court drama that flourished during Elizabeth's reign that such changes in the aesthetics of display became particularly apparent. Within Shakespeare's career, most notably, one can see the interdependence of the queen's two bodies give way, following Elizabeth's death, to an increasing emphasis on the metaphysical nature of the crown over and above the individual monarch who momentarily held sway over the land.

Hamlet is one of the plays to appear during the time when people were finding it necessary to revise the aesthetics of Elizabethan display to suit an impending Jacobean reality. The play presents two quite different displays of power, each centered around the body of a different aristocratic female. On the fate of Gertrude and the disposition of her body depends both the well-being of the state and the fate of the royal Danish line. In this sense, Gertrude belongs among Elizabethan representations of the queen's two bodies. These characteristically equate the health of the state with that of the queen's body. In having Hamlet stage the play within the play, however, Shakespeare uses the aristocratic female body in a different way. The Player Queen behaves like all the other aristocratic females on the Jacobean stage who are tortured, stabbed, poisoned, or hung. It is by defining them as the site of pollution and removing them from the line of authority that patriarchal power is itself authorized.

To read *Hamlet* historically, in my opinion, it is not our task to explain away this about-face in the strategies of political display. This is what we do whenever we try to contain the contradictions posed by the two queens within Hamlet's "consciousness," making them his problem rather than our own. Contrary to this way of reading *Hamlet*, I would like to consider the play as a refiguring of the monarch's body in view of Elizabeth's immanent decay. Far from embodying the power of the state itself, the aristocratic woman would, in the years immediately following the production of *Hamlet*, provide playwright after playwright with a figure for the source of pollution. As such, she was none the less subject to the aesthetics of display, for her purification alone appeared to insure the perpetuity of power.

III

I shall offer a brief description of Lavinia from the much abused play *Titus Andronicus* by way of background against which we can determine the ways in which Gertrude observes the Elizabethan formula. Despite its popularity on the Elizabethan stage, *Titus* generally strikes modern readers as a thoroughly debased representation of sexuality (Brooke 1968: 13–15).[3] Yet what may appear as perverse and gratuitously violent assumes this form because the dramatic action of the play turns on the whole notion of the State as the body of an aristocratic female. Particularly disturbing is the fact that Titus' daughter is not only raped and disfigured in the second act of the play but also brought upon the stage to display her mutilated condition. The sheer spectacle of a woman, herself dismembered, carrying her father's amputated hand in her mouth has not earned this play a particularly high place in a literary canon based on lofty ideals and good taste. The mutilation of Lavinia's body has been written off by critics as one of the exuberant excesses of an immature playwright or else as the corrupting influence of another poet. But I find it more useful for the purposes of historicizing Shakespeare to consider these sensational features as part of a political iconography available to the playwright, one which he felt obliged to use as well as free to exploit for his own dramatic purposes. With this purpose in mind, then, we can consider as culturally important information the otherwise outrageous scene in which Titus receives his own hand along with the heads of his two sons from Saturninus, the emperor. Seeing the human members which have been severed from himself, Titus issues this memorably gruesome command,

> Come, brother, take a head,
> And in this hand the other will I bear;
> And, Lavinia, thou shalt be employed;
> Bear thou my hand, sweet wench, between thy teeth.
> (III.i.279–82)

To tell her father she has been raped as well as mutilated, Lavinia has to rifle through a volume of Ovid with her handless arms until she finds the account of Philomel. Shakespeare's stage direction reads, "She takes the staff in her mouth, and guides it with her stumps, and writes" (s.d., IV.i.78). What is important in this – as in the other scenes where Lavinia's body appears as synecdoche and emblem of the disorder of things – is that Shakespeare has us see the rape of Lavinia as the definitive instance of dismemberment.

I say this knowing that it defies the logic inherent in the figure of rape. We are accustomed to think of rape as a forcible violation of some sacred cultural boundary enclosing the aristocratic body if not that of the private individual. But Elizabethan drama does not use rape in this way. Lavinia's rape represents the crime of dismemberment. The mutilation of Lavinia's body restates her father's self-inflicted amputation, his dicing up of the emperor's stepsons for their mother's consumption, and all the slicing, dicing, chopping, and lopping that heaps bodies upon the stage in *Titus Andronicus*. Lavinia's body encapsulates and interprets this seemingly gratuitous carnage in a way that must have been clear to an Elizabethan audience, for her body was that of a daughter of the popular candidate for emperor of Rome, the first choice of wife for the emperor of Rome, and the betrothed of the emperor's younger brother. That as such she stands for the entire aristocratic body is made clear when Marcus Andronicus, inspired by the pile of bodies heaped before the banquet table, enjoins the citizens of Rome, "Let me teach you how to knit again/ . . . These broken limbs into one body" (V.iii.70–2).

The logic of dismemberment is not that different if one is willing to consider it as such. Dismemberment entails the loss of members. Thus the initial gesture of penetration is not so well pronounced in Shakespeare's version of the Philomel story as the mutilated condition of Lavinia's body which both conceals and points back to the act of rape. Rather than the object of illicit lust, Lavinia's body provides the setting for political rivalry among the various families with competing claims to power over Rome. For one of them to possess her is for that family to display its power over the rest –

nothing more nor less than that. By the same token, to wound Lavinia is to wound oneself, as if dismembering her body were dismembering a body of which one were a part, and thus to cut oneself off from that body. It is in pursuing this logic that one sees how Titus' farewell to Lavinia transforms the concepts of dishonor and pollution usually associated with rape into quite a different order of transgression: ''Die, die, Lavinia, and thy shame with thee,/ And with thy shame thy father's sorrow die'' (V.iii.41–2). True to this suggestion that she is his body because she is the body of Rome, the play demonstrates that the murder of Lavinia is a self-inflicted wound on Titus' part. It leads to the death of the entire ruling body, competing families and all.

IV

Gertrude's body observes the same political imperative as that of Lavinia and the young Elizabeth. To possess her body is to possess the State. So powerful is the queen's body that it takes precedence over the laws of primogeniture allowing Claudius to rule instead of Hamlet. According to common practices of primogeniture, when Old Denmark died his crown and his land should have passed uncontested to his son. Had he died without issue, then and only then could the crown pass to his brother. But Claudius's ascendence does not observe this principle of inheritance. His claim to power – and the election that ratified his claim – rested on his claim to Gertrude's body. It is Claudius's acquisition of power through his marriage to Gertrude that gives rise to the dilemma organizing this play, the action of which turns upon the meaning and disposition of Gertrude's body.

There is a logic at work in *Hamlet* which explains this source of power. Such a logic, however, is neither to be found in the Danish history of Saxo Grammaticus which Shakespeare consulted as a source for his play, nor can we attribute it to the peculiarities of Danish political practice with which Shakespeare may or may not have been familiar. We can, however, see this logic at work in any number of romantic comedies where a young man comes to possess power, wealth, and land through marriage. Bassanio's marriage to Portia, Lorenzo's to Jessica, Petruchio's to Kate, and Sebastian's to Olivia are cases in point. In each play the female provides access to a patrimony that belongs to another male. The patrimonies thus in question might range from the kingdom of Belmont to something considerably less grand, like the dowry Kate brings with her to the marriage. I am suggesting that when he has Claudius come to the throne through marriage to Gertrude Shakespeare is drawing on the same theory of power that also organized his romantic comedies.

Hamlet's obsession with his mother's body can be explained in terms of this theory. When Hamlet urges his mother to refrain from having sexual contact with Claudius, his words, taken at face value, quite accurately describe the problem that arises from the premise that power inheres in her body. He represents the queen's coupling with Claudius as the gratification of a monstrous appetite:

> Here is your husband, like a mildewed ear,
> Blasting his wholesome brother. Have you eyes?
> Could you on this fair mountain leave to feed,
> And batten on this moor? (III.iv.64–7)

To think of Gertrude's union with Claudius as a form of gorging (''batten'') makes sense only if one remembers that she should represent the aristocratic body itself. Hamlet's language transforms the ideal representation of that body, the body of the queen, into one that is quite grotesque and common. In the passage quoted above she has become the voracious mass body that regularly stands opposed to the aristocratic principles of exclusion and hierarchy.[4] And lest we miss the point, Shakespeare has Hamlet elaborate this view of the queen's body. To mate with Claudius, in his words, is:

> but to live
> In the rank sweat of an enseamed bed,
> Stew'd in corruption, honeying and making love
> Over the nasty sty. (III.iv.92–5)

But how does Gertrude become this gorging, sweating, corrupt, and bestial woman? The answer seems to lie with Hamlet's description of the man he thinks wrongly possesses her body. Thus Hamlet describes Claudius as

> a Vice of kings,
> A cutpurse of the empire and the rule,
> That from a shelf the precious diadem stole,
> And put it in his pocket . . .
> A king of shreds and patches — (III.iv.98–102)

As he leaves, Hamlet implores his mother not to let ''the bloat king tempt you again to bed'' (III.iv.182). To have the prince call Claudius ''a bloat king,'' a lecher, a ''cutpurse of the empire,'' ''a Vice of kings'' is for Shakespeare to construct the usurper out of the same materials he used in fabricating Falstaff. Unlike Claudius, then, for whom Gertrude's body is Denmark, Hamlet understands Gertrude's body as the vessel through which the royal blood of the Danish line has passed; she is not the political body incarnate. She disrupts the continuity of Old Denmark's line by authorizing his brother's usurpation of the throne. That she does so certainly suggests that the conditions of the queen's body and that of the State are not the same. This was a mistake for which Hamlet played out a tragic fate, but it was nevertheless a possibility Shakespeare could imagine and through which create new dramatic uses for the aristocratic female.

V

By the end of the 1590s, the physical condition of the heirless queen evidently made it necessary to reconsider the relationship between that body and the political strength of England. Although a concerted effort was made to maintain her traditional hold on the popular imagination, the queen's age made it necessary to modify the displays which had identified her natural body with the power inhering in the body politic. Aware that her health was increasingly a matter of political gravity, Elizabeth sought ways to insist upon the vitality of her body. During the Christmas celebrations of 1600, for example, Elizabeth made a public show of dancing with Duke Bracciano. John Chamberlain writes, "The Queen entertained him very graciously, and to show that she is not so old as some would have her danced both measures and galliards in his presence" (1939: 115). Despite her attempts to comply with the aesthetics of display, the signs of her age were everywhere to be seen. At the opening of parliament in 1601, it was reported, "her robes of velvet and ermine had proved too heavy for her; on the steps of the throne she had staggered and was only saved from falling by the peer who stood nearest catching her in his arms" (Jenkins 1958: 321). Her increasing feebleness threatened to shake the political foundations of the State. When, in August 1599, Londoners lived in fear of a Spanish invasion, John Chamberlain explained to Dudley Carleton how the appearance of military commanders at the Paul's Cross sermons was regarded by the crowd:

> The Lord General with all great officers of the field came in great bravery to Paul's Cross on Sunday . . . and then was the alarm at the hottest that the Spaniards were at Brest . . . .
>
> The vulgar sort cannot be persuaded that there was some great mystery in the assembling of these forces, and because they cannot find the reason for it, make many wild conjectures and cast beyond the moon: as sometimes that the Queen was dangerously sick. (1939: 83)

Rather than the routine attendance of military men to hear a sermon the "vulgar sort" took the military presence to mean that the queen was certainly failing. With the failure of the monarch's natural body, they assumed that the magical power of the crown was also in question, and the nation, therefore, in a state of imminent peril.

What is most important for purposes of my argument, however, is the suggestion that Elizabeth's age made it dangerous to equate her body with the body politic. We might understand the Essex rebellion in relationship to new fault lines in the iconography of power that further threatened the stability of the Tudor reign, shakiness that consequently afflicted the reigning aesthetics of monarchy. Angry at the queen for her support of Cecil, angry at her, too, for reprimanding him when he granted wholesale knighthoods in Ireland, angry at being denied the opportunity to dispense patronage in

England, and angry at the recent Star Chamber proceedings against him, Essex is said by Camden to have complained bitterly that Elizabeth was "grown an old woman and as crooked in mind as in her carcase" (Birch 1970: 463). Clearly Essex believed that the symbolic powers of the queen's body were susceptible to appropriation. Even after the government discovered his plans, Essex behaved as if the mere display of his colors and the support of relatives, friends, clients, and household retainers would give him the authority he needed to overrule the queen. Essex no doubt assumed that the queen's body contained the magical power of the blood, but evidently he did not see that magic as the sole source of English political power. Indeed, in his manner of using the aesthetics of display to rebel against the queen's authority, he distinguished between the immanent magic of blood and the queen's symbolic display of that power, as if to say that such a display of power could empower him as well. If he behaved like a monarch, according to this inverted logic, he could attract support from the people. Following his arrest, the indictments charged Essex with attempting "to usurp the Crown," and the Earls of Essex, Southampton, Rutland, and Sandys with conspiring to depose and slay the queen (Akrigg 1968: 120–1). Two days after his conviction, Essex contested this change, claiming that he meant only to seize the queen and use her authority to change the government. He did not want to weaken her authority but merely to remove her advisors and condemn them for mismanaging the state (Akrigg 1968: 127). In either case, however, he had questioned the bond between the monarch's two bodies. Whether he intended to overthrow the queen – which is unlikely – or simply to force her to name the successor of his choice, Essex had granted the display of power priority over the natural body of Elizabeth and, by implication, over the mystic line of succession.

The question of which of the two had priority – the natural body or the metaphysical body of the monarch – held little fascination for people during much of Elizabeth's reign, so firmly linked was the national identity with her figure. For this very reason, however, the question became all the more urgent with the approach of the queen's death. On the one hand, her death meant the end of the only English monarch most of the population had ever known and in whom they read the fate of the nation. To detach the whole idea of state authority from the queen's body was – as John Chamberlain's report of the panic at the Paul's Cross sermon suggests – a dangerous proposition. On the other hand, it was just as dangerous to maintain the iconic link between that body and the state, for the aged virgin bore not only signs of decay but also signs of sterility that told of an uncertain future for England. Legitimate power, as the Essex rebellion suggested, might pass to whomever put on the symbolic displays that legitimized power. Thus we may speculate that it became necessary for playwrights to stress the metaphysical over the natural body of the monarch – to demystify the queen's body, that is, and to remystify patrilineage.

VI

The Murder of Gonzago – Hamlet's play within the play – is but one of a number of performances that would bring about this historical transformation. No longer iconic, the Player Queen's body opens up the possibility of another poetics of power, one that would come to dominate the stage once Elizabeth passed from the scene. Though the female body in Hamlet's drama is no longer wed to the land, it nevertheless authorizes monarchy. In fact, as I have suggested, it must be detached from the land in order to authorize monarchy. If we look ahead two or three years to the Jacobean aesthetics that come to dominate the public theater, we see that numerous playwrights find it necessary to torture, smother, strangle, stab, or poison an aristocratic woman in order to stage a tragedy. Desdemona, Cleopatra, Goneril, Regan, Cordelia, the Duchess of Malfi, and Vittoria Corombona are among those who testify to this compulsion. These aristocratic females all share one of two features in common: (1) they are either the subject of clandestine desire or (2) they are the object of desire that threatens the aristocratic community's boundaries. Their innocence notwithstanding, women in *The Revenger's Tragedy* must be poisoned once they become objects of adulterous desires. The Count Montsurrey in *Bussy D'Ambois* tortures Tamyra for her secret assignations, and Othello murders Desdemona because he assumes that she has been guilty of infidelity. True, he is wrong to doubt her innocence. On the other hand, Desdemona has, like the Duchess of Malfi, violated the law of her blood in marrying him. And her marriage to the Moor echoes the mismating of the Egyptian and the Roman, of the duchess and her steward, of the duke and the white devil, as well as those two queens of Britain who lust for the bastard Edmund. In each case, these women are subjected to spectacular scenes of punishment, because each poses a threat to the Jacobean notion of monarchy.

In Jacobean tragedy, the line between the two social bodies – the aristocratic body and that of the people – appears to close. Othello, Malfi's husband, Vittoria Corombona, and Edmund seem capable of becoming part of the aristocratic body, but the fact of their transgression is acknowledged as they produce disease, filth, and obscenity that must be purged in order for there to be a pure community of aristocratic blood. The staging of such scenes of punishment are attempts to rid the community of some kind of pollution, the conditions for which Mary Douglas describes:

> Pollution is a type of danger which is not likely to occur except where the lines of structure, cosmic or social, are clearly defined . . . a polluting person is always in the wrong. He has developed some wrong condition or simply crossed some line which should not have been crossed. . . . Pollution can be committed unintentionally, but intention is irrelevant to its effects. It is more likely to happen inadvertently. (1978: 113)

It is according to this sort of logic that innocent characters are often slaughtered on the stage while morally despicable conduct seems to go unpunished.

In Jacobean tragedy, any sign of permeability poses a threat to the community. The bodies of such women as Desdemona, Tamyra, the Duchess of Malfi, and Lear's daughters blur the distinction between what belongs inside and what must be kept outside the aristocratic community. Each of these women negotiates a sexual union that is represented as a form of pollution. Although the female body must be understood in terms of the metaphysics of blood, on the Jacobean stage the female body no longer exists in the same iconic relationship with that of the monarch and with the magical power of blood. If anything, Jacobean tragedy insists on this disruption of the Elizabethan model all the more forcefully by imagining the state as nothing else but the blood, the blood in its purest form; in other words, the blood of the patriarch. A closer look at one of the more famous scenes of punishment will reveal the logic that governs pollution and purification rituals as opposed to that of dismemberment. Although Renaissance drama consistently inflicted elaborate and brutal punishments on the bodies of aristocratic women, differing techniques of mutilation reveal underlying political tendencies that changed the aesthetics and theatrical display with the change in monarchs.

As he tortures his wife in *Bussy D'Ambois*, the Count Montsurrey claims it is not he but her lust that murders her:

> The chain shot of thy lust is aloft
> And it must murder; 'tis thine own dear twin. (V.i.91–2)

Lust doubles the woman. In that it produces either a desirous or a desiring self, lust makes her monstrous in some way. It should not be surprising in this regard to discover that twinning and doubling occupied a major section of Renaissance books on monsters and monstrosity. In his book on monsters and marvels, Ambroise Paré describes twins who share a single head, twins joined at the belly, or twins that have but a single anus between them (1982: 27).[5] Or he describes the monstrous as a single figure with twinned arms on one side of the body, one that has double the number of legs, another with extra fingers, and those bearing extra members of other kinds as well. Particularly important among these is the hermaphrodite. By virtue of possessing a second set of sexual organs, the hermaphrodite resembles other monsters in that he violates natural categories. In doing so, however, the hermaphrodite could also be used to clarify these differences. It was always necessary to determine which set of sexual organs was dominant and thereby remove the hermaphrodite from the status of a monstrosity. Paré explains that whenever both sets of sexual organs were fully formed in an individual, both ancient and modern law obliged such monsters to say "which they wish to use, and then they are forbidden upon pain of death to use any but those they have chosen" (ibid.). By containing an extra member, hermaphrodites not only violated the natural order, as did all Paré's monstrous creatures, but they also threatened to pollute the community. It was therefore necessary to suppress the supplementary feature.

The monstrous woman also possesses an extra member. Women subjected to punishment are those who either bring an extra member into the body politic or else take on the features of masculine desire themselves. Malfi's marriage adds an extra member to the aristocratic body, and this member is referred to as the "hermaphrodite." Claiming she has polluted his blood, Malfi's brother takes the form of a lycanthrope. Similarly monstrous, Vittoria Corombona is a masculine woman, twinned by lust, and rendered so monstrous by desire that she is called the White Devil. Indeed, it seems that whenever the rigid boundaries that define the pure community are obscured within the female body, Jacobean drama reclassifies the woman as a monster, suggesting, Mary Douglas argues, that pollution represents a type of danger that occurs where clearly defined social lines have been muddled. The bodies of such women as Tamyra, the Duchess of Malfi, Desdemona, Cleopatra, Goneril and Regan, and Vittoria Corombona, among others, obscure that boundary distinguishing what may be contained inside the community from what must be kept out. Thus their bodies, like that of the hermaphrodite's, provide the place where difference must be re-established.

Having noted then how the body of the aristocratic female on the Jacobean stage takes on a decidedly different figural connotation, we can consider how the particular form of sexual violence to her differs from that portrayed in an Elizabethan play like *Titus Andronicus*. To begin formulating an answer, I would like to return to one of the most self-conscious of the Jacobean plays and examine its dramatization of punishment. Here are the terms in which Tamyra would have us understand the forthcoming scene of her torture in *Bussy D'Ambois*:

> Hide in some gloomy dungeon my loathed face,
> . . .
> Hang me in chains, and let me eat those arms
> That have offended: bind me face to face
> To some dead woman taken from the cart of
> Execution (V.i.103–10)

This passage sets up a parallel between the husband–wife relationship and that of sovereign and subject that would have automatically made sense of Tamyra's crime as well as the scene of punishment: the wife's assertion of power against her husband must be understood in relation to the subject's assault upon the sovereign's power. In the second chapter of *Discipline and Punish*, Foucault suggests that, in considering such scenes in pre-Enlightenment culture, we must take the homology literally. Whatever attacks the law of the sovereign also attacks him physically, since the force of the law is the force of the prince. When popular power – always expressed in the language of festival and always illegitimate – forgets the truth that legitimate power is absolute, according to Foucault's model, scenes are staged to display that radical disymmetry of power. Foucault writes:

> If torture was so strongly embedded in legal practice, it was because it revealed truth and showed the operation of power. It assured the articulation of the written on the oral, the secret on the public, the procedure of investigation on the operation of confession; it made it possible to reproduce the crime on the visible body of the criminal; in the same horror, the crime had to be manifested and annulled. The nature of the threat posed by the criminal can only be intensified when the crime is a crime against the aristocratic body itself. (1977: 50)

The wife's crime against her husband *is* a crime against the crown. The punishment of unchaste aristocratic women therefore displays the truth of the subject's relation to the state. It displays the disymmetrical relationship by imprinting the crime on the subject's body, in this way demonstrating the state's absolute power over that body. Tamyra's husband orders her to write the name of the go-between as she is stretched on the rack. At the same time, he repeatedly stabs her arms. Since the permeability of her body wounded him, the cultural logic which organizes the play dictates that he should cut openings upon her, for this makes her crime against the state legible. As they subordinated female to male in such an extravagantly artificial manner, dramatists testified to the absolute power of the state. But it is important to understand that certainly in England this subordination could only take place once the figure of the aristocratic female body was understood as something separate from the crown. Given my construction of the cultural milieu in which Shakespeare wrote, Hamlet's attempt at staging a play is very much an attempt on the playwright's part to imagine a situation in which political power was not associated with a female and the aristocratic female was not iconically bonded to the land.

VII

The Murder of Gonzago is Hamlet's attempt to locate and purge a corrupt element within the aristocratic body. In this respect, he acts in his capacity as would-be sovereign, for Shakespeare gives to Hamlet the sovereign's power to discover and punish a crime against the aristocratic body. As he explains:

> I'll have these players
> Play something like the murther of my father
> Before mine uncle. I'll observe his looks,
> I'll tent him to the quick. If a' do blench,
> I know my course. (II.ii.594–8)

Hamlet means the play to ''tent,'' or probe, Claudius as with a dagger that opens an infected wound. In this way, he intends to re-enact the crime of regicide for the purpose of punishing the murderer. By staging a torture and seeking to extract a confession, Hamlet takes it upon himself to exercise what Foucault has claimed was ''the absolute right and the exclusive power of the sovereign.'' To carry the point still further, chronicle history plays (which were, along with romantic comedies, a preferred Elizabethan mode of drama) establish the legitimacy of a monarch both by staging his control over truth through the exercise of punishment and by displaying his ability to possess the territory of England by means of a miraculous victory over a stronger opponent. A year or two into the reign of James I, romantic comedies and chronicle histories were eclipsed by problem comedies and tragedies that threatened if they did not actually enact extravagant scenes of punishment on the body of an aristocratic woman. This, rather than the monarch's possession of the land, established his (always his) legitimacy.

But Hamlet's play fails in two respects to materialize as a spectacle of punishment which would establish Hamlet's power over Claudius. Because the play is only a play, first of all, and not an official ritual of state, its truth is marked as a supposition rather than a re-enactment of the truth. It is another instance of Shakespeare's giving Hamlet a mode of speech that cannot constitute political action because it automatically translates all action onto the purely symbolic plane of thought and art. Only here, as opposed to earlier moments in the play where Hamlet's speech renders him unable to act, it is not his use of Stoic discourse but of the Senecan mode of tragedy that turns the exercise of power into a purely symbolic gesture. Secondly, even as a symbolic gesture, the play fails to hit its mark. Hamlet has chosen to produce *The Murder of Gonzago* to display a political truth. The ''play,'' he says, ''is something like the murder of my father.'' Indeed, the play is a re-enactment of the fratricide in that it portrays the aristocratic body (one brother) turning against itself (another brother) to inflict a wound that will ultimately kill them both. But Hamlet's gloss on the play informs us, curiously enough, that he has chosen a play portraying the murder of an uncle by his nephew. Hamlet explains, ''This is one Lucianus, nephew to the king'' (III.ii.244), and then adds:

> 'A poisons him i' th' garden for his estate. His name's Gonzago, the story is extant, and written in very choice Italian. You shall see anon how the murtherer gets the love of Gonzago's wife. (III.ii.261–4)

To say that this play shows ''how the murtherer gets the love of Gonzago's wife'' is to say that *The Murder of Gonzago* does not depict regicide as a crime against a patrilineal system of descent. The same thing is further indicated by the fact that it is not first and second sons, or even the progeny of siblings, who contend for the throne, although this is initially the order of conflict that brought down old Hamlet and provides the framework of

Shakespeare's play. Unlike Claudius, who murders his brother, Hamlet's Gonzago murders his uncle, and it is through possession of his uncle's wife that this usurper symbolically gains control of Gonzago's patrimony. That Hamlet represents the original fratricide as the murder of an uncle by a nephew has empowered many modern readers to regard the play within the play as psychological information – as representing, that is, Hamlet's wish to possess his mother and not as Hamlet's attempt to reveal the guilt of Claudius in the re-enactment of that crime as punishment. But given a cultural milieu in which any display of the aristocratic body would have been highly meaningful politically, I think it more likely that, by casting the murderer as nephew to the duke, Shakespeare deviated from his source and from the kinship relations dominating the play as a whole in order to represent the queen's body as an illegitimate source of political authority (Bullough 1973: 30, 172–6). It is highly doubtful that Shakespeare meant to say both things – psychological and political – at once, since they more or less cancel each other out. Hamlet cannot be desiring his mother (according to the modern Oedipal pattern) and still want to identify her as the site of political corruption and danger to the state.

This revision of Elizabethan thinking is more likely Shakespeare's attempt at updating Claudius's crime to address historical circumstances very different from those in which he staged the chronicle history plays. Having dramatized how power passes into the wrong hands through the body of a woman, however, *Hamlet* folds back into an Elizabethan mode of thinking and equates Hamlet's abortive attempt to enact the rites of punishment with Claudius's crime. Both assault the sovereign's body rather than establish the absolute power of the aristocratic body over that of its subject as both turn out to be self-inflicted wounds. The play concludes according to the Elizabethan logic which governs *Titus Andronicus* by heaping up the bodies of the royal family where there should have been a banquet scene. Thus this play materializes the truth that the murder of one member of the aristocracy by another is an assault on the entire body or, in other words, an act of suicide.

I am suggesting that the dilemma of the play arises from and turns upon the meaning and disposition of Gertrude's body. Where Lavinia provided the site for the various forces competing for Rome, Gertrude's body stages a conceptual shift in the representation of political disorder. Her body becomes the place where the iconic bonding of blood and territory breaks down into competing bases of political authority. Claudius's authority rests on his marriage to Gertrude. To Hamlet, on the other hand, authority depends on birth. The question is not a matter of which family embodies legitimate power over the land, but a matter of which claim – blood or the possession of territory – is more important in constituting legitimate authority.

The play within the play represents the Player Queen's body in terms that go still further in contradicting the politics of the body that governed many of the symbolic practices of an Elizabethan England. But *The Murder of Gonzago* ends before the logic of the representation can play itself out. Even so, the dumb show which precedes the performance – in combination with Hamlet's gloss and his uncle's angry reaction to it – bears intimations of another politics of the body. Hamlet's gloss on the play tells of a nephew poisoning his uncle, the king, and then wooing the queen. The audience has seen the king murdered "for his estate," Hamlet explains, only to add, "You shall see anon how the murtherer gets the love of Gonzago's wife" (III.ii.261–4). After this statement, Claudius rises and closes down the theater before the drama of illicit sexual relations can fairly get under way. Coming when it does, this interruption of the play within the play further distinguishes two acts of treason – the seizure of royal property and the possession of the queen's body – one from the other. It is more than a little interesting to note that the threat to the aristocratic body is a double threat which distinguishes two points where the aristocratic body could receive a mortal wound. It might be said that Shakespeare formally posed a political threat of this same magnitude simply by mutilating the female body, but now he feels somehow compelled to launch two separate assaults: one, to lose the land and, two, to destroy the sacred symbols of state. To reinforce the sense that these sources of power are separate and distinct, Shakespeare has the Player Queen in Hamlet's script describe her own sexual behavior as an assault on the political body separate and distinct from that which destroyed the king's natural body: "A second time I kill my husband dead,/ When second husband kisses me in bed" (III.ii.184–5).

I am certainly not suggesting that such splitting of the political body according to sex makes sexual desire any less political than it was in earlier drama. It only means – in *Hamlet* at least – that the politics of the body is susceptible to change. The Player Queen in Hamlet's revised script opens up a new category of crime. Her body is no longer the state. Or perhaps it is more accurate to say this the other way around, that the political body is no longer a woman. Accordingly, the Player Queen ceases to be a source of legitimate power. Like Gertrude, she crowns a counterfeit monarch who possesses the land on a basis other than blood. In the play within the play, then, the female body becomes a place where the body politic can be corrupted. And as the Player Queen corrupts rather than legitimates the blood, she corrupts the official iconography of state; she becomes an object of desire in her own right, a desire for the signs and symbols of power dissociated from the metaphysics of blood. This presumably lies behind Hamlet's promise that the bulk of the play will dramatize "how the murtherer gets the love of Gonzago's wife" (III.ii.363). Different conceptions of the female body thus interact in *Hamlet* to the external fascination of modern readers. It is particularly difficult for us to understand the political basis of this debate because doing so depends upon seeing the body politic as female. The second notion of the female body develops with the revision of the first under the pressure from Elizabeth's aging body and the subsequent formation of new political resistance to the patriarchal ideal.

This second notion of the body politic sees the female body – and by this I mean specifically that of the aristocratic female – as the symbol and point of access to legitimate authority, thus as the potential substitute for blood and basis for counterfeiting power. The historically earlier view sees the aristocratic female body and the political body as one and the same, a view that resists our attempts at privileging sexual differences over those based on blood. The second view of the body allows us to translate political relationships into sexual relationships as it cuts a clear difference between the body politic and that of the female. With the possibility that her body serves as the symbolic substitute for some original body, furthermore, comes the possibility of construing the aristocratic woman as an object of sexual desire rather than as the means to political power. But to regard Gertrude in the light of modern sexuality is to reverse the priorities of Jacobean thinking where sexual desire always has a political meaning and objective. We regularly perform this gesture of historical reversal when we read the political formations overlapping in *Hamlet* as events in an interpsychic melodrama where the two queens are no longer political figures but ciphers of Hamlet's relation to his mother. Instead, I am stressing the figural discontinuity between Gertrude's body and that of the Player Queen.

The fate of Gertrude makes *Hamlet* an Elizabethan play. Upon the condition of her body depends the health of the state. Like Old Denmark before her, Gertrude dies from taking poison into her body, and the same poison strikes down Hamlet, Laertes, and Claudius as well. Her death thus initiates the heaping of bodies which characterizes the Elizabethan Seneca; the wounding of one of its members is the wounding of the entire political body. In this case, however, the infiltration of that body with poison puts an end to the Danish line. This is the fate, experienced by one, that all members share. By this means, and not by a blow, then, is how Shakespeare imagines a lethal assault on the body politic. But *The Murder of Gonzago* takes the logic of this figure one step further. Shakespeare uses poison to threaten the political body in a manner which appears to contradict the politics of dismemberment. Merely by inciting desire, the queen's sexuality becomes a form of corruption equivalent to but not the same as the poison which has been poured into the ear of the sleeping king. Thus Hamlet insists upon shifting the crime from the fact of regicide to the act of "incest," the term which he uses to describe the relationship between Claudius and Gertrude, his brother's wife.

The play within the play can be viewed as Hamlet's way of distinguishing the one crime from the other, a distinction which relocates the political body in the female. Such a shift can occur only if the queen does not embody patriarchal power, otherwise any misuse of her sexual body, however deliberate on her part, would constitute a direct assault on the whole concept of patriarchy. Hamlet's obsession with the misuse of the queen's sexuality, more than with his uncle's possession of the state, transforms the threat of dismemberment into one of pollution. We might say that, in redefining the nature of the threat against the body politic, Hamlet attempts to stage a Jacobean tragedy. But the political context of Hamlet's play – in other words, Shakespeare's play – proves more powerful than Hamlet's attempt to transform the political relationships that prevail outside his theater. It is to the Elizabethan dynamic of competition that he eventually succumbs as Shakespeare brings Hamlet's struggle on behalf of a later construction of patriarchy to an Elizabethan conclusion. Hamlet fails to transform the iconography of state. In the tragedies that follow *Hamlet*, however, the aristocratic female is regularly caught up in assaults on the principle of blood only to be tortured or murdered, thus testifying to the metaphysics of the monarch's body.

Notes

1 The anthropologist Emily Martin concludes after studying representations of the body in medical school textbooks, magazines, journals, and newspapers that everything from cell structure to reproduction bears features of the modern industrial state and, as such, shapes our most basic evaluations of ourselves and others. Cells are described as factories, the brain is a co-ordinating center for sending and receiving messages, the AIDS virus is represented as a factory producing anti-immune tanks. If anything, Martin's study (1987) tells us the simple truth that the body is always a product of a particular culture.
2 For discussions of the politics of the queen's display of her body, see Greenblatt (1980: 166–8), Goldberg (1983: 28–30), and Bergeron (1971: 12–32).
3 Despite the persistently low esteem in which critics hold *Titus*, there have been useful efforts to explain the aesthetics of violent display in the play (Bevington 1984: 29–32).
4 Of the mass body and its opposition to what I am calling the aristocratic body, Bakhtin writes that "it is outside of and contrary to all existing forms of the coercive, socioeconomic and political organization" (1968: 255). Contrary to the closed, rigidly hierarchized, and pure figure of the desirable woman which authorized the blood, the grotesque body is open, heterogeneous, undifferentiated, sensual, concrete, endlessly copulating, always hungry, and forever reproducing itself. I am also indebted to Peter Stallybrass and Allon White for their important discussion of the mass body.
5 I am indebted to Stephen J. Greenblatt for calling this material to my attention. He kindly allowed me to consult a manuscript "Fiction and Friction in *Twelfth Night*" (1988: 66–93) in which he discusses at length the Renaissance preoccupation with monstrous births.

References

Akrigg, G.P.V. (1968) *Shakespeare and the Earl of Southampton*. Cambridge, Mass.: Harvard University Press.
Armstrong, Nancy (1988) "The Gender Bind: Women and the Disciplines." *Genders* 3 (Fall): 1–23.

Axton, Marie (1977) *The Queen's Two Bodies: Drama and the Elizabethan Succession*. London: Royal Historical Society.

Bakhtin, M.M. (1968) *Rabelais and His World*. Trans. Hélène Iswolsky. Cambridge: MIT Press.

Bergeron, David (1971) *English Civic Pageantry, 1558–1642*. Columbia: University of South Carolina Press.

Bevington, David (1984) *Action is Eloquence: Shakespeare's Language of Gesture*. Cambridge, Mass.: Harvard University Press.

Birch, Thomas (1970) *Memoirs of the Reign of Queen Elizabeth*, 2. [1754]. New York: AMS.

Brooke, Nicholas (1968) *Shakespeare's Early Tragedies*. London: Methuen.

Bullough, Geoffrey (1973) *Narrative and Dramatic Sources of Shakespeare*, 7. New York: Columbia University Press.

Chamberlain, John (1939) *The Letters of John Chamberlain*, 1. Ed. Norman McClure. Philadelphia: American Philosophical Society.

Chapman, George (1976) *Bussy D'Ambois. Drama of the English Renaissance*, 2. Ed. Russell Fraser and Norman Rabkin. New York: Macmillan.

Comaroff, Jean (1985) *Body of Power, Spirit of Resistance: The Culture and History of a South African People*. Chicago: University of Chicago Press.

Douglas, Mary (1978) *Purity and Danger: An Analysis of Concepts of Pollution and Taboo*. London: Routledge & Kegan Paul.

Eccles, Audrey (1976) *Obstetrics and Gynaecology in Tudor and Stuart England*. Kent, Ohio: Kent State University Press.

Foucault, Michel (1977) *Discipline and Punish: The Birth of the Prison*. Trans. Alan Sheridan. New York: Vintage.

Goldberg, Jonathan (1983) *James I and the Politics of Literature: Jonson, Shakespeare, Donne and their Contemporaries*. Baltimore: Johns Hopkins University Press.

Greenblatt, Stephen J. (1980) *Renaissance Self-Fashioning: From More to Shakespeare*. Chicago: University of Chicago Press.

——— (1988) *Shakespearean Negotiations: The Circulation of Social Energy in Renaissance England*. Berkeley: University of California Press.

Jenkins, Elizabeth (1958) *Elizabeth the Great*. New York: Harcourt Brace.

Laqueur, Thomas (1986) "Orgasm, Generation, and the Politics of Reproductive Biology." *Representations* 14: 1–41.

Martin, Emily (1987) *The Woman in the Body: A Cultural Analysis of Reproduction*. Boston: Beacon.

Paré, Ambroise (1982) *On Monsters and Marvels*. Trans. Janis L. Pallister. Chicago: University of Chicago Press.

Phillips, John (1973) *The Reformation of Images: Destruction of Art in England 1535–1660*. Berkeley: University of California Press.

Shakespeare, William (1972) *The Riverside Shakespeare*. Ed. G. Blakemore Evans. Boston: Houghton Mifflin.

Stallybrass, Peter and White, Allon (1986) *The Politics and Poetics of Transgression*. Ithaca, NY: Cornell University Press.

Strong, Roy (1963) *Portraits of Queen Elizabeth I*. Oxford: Clarendon.

4

The other Quixote

George Mariscal

I

Violence, Power, Ideology, Discourse: these are words that have enchanted literary scholars in recent years. Taken together, they constitute a field of critical practice which, in its rejection of formalizing of any type, proposes to redefine the relationship between text and context. As with all methods, however, there are potential dangers, not the least of which in this case (and not unlike the one to which Sancho Panza's beloved proverbs fell prey) is that we essentialize such terms, transforming them into unassailable giants who indiscriminately stride over diverse situations and epochs, and force us to forget the concrete historical realities to which they refer. Yet despite such dangers, can there still be any doubt that literature and culture in general are inseparable from such concerns? For not only do literary texts often represent and reveal the workings of violence, power, and the like, but these same texts (and their grouping together as canon) are always conceived and produced by an economy of exclusion, conflict, and rivalry itself inextricably caught up in a wider network of social interests and relations. Today, even the classics and masterpieces can no longer be safely hidden away from history like wineskins tightly filled with transcendental truths.

Nine years after the publication of Cervantes' 1605 *Don Quixote*, a second novel appeared by a certain Alonso Fernández de Avellaneda purporting to be a continuation of the misadventures of the Manchegan knight. In his prologue, Avellaneda launches a scathing personal attack against Cervantes, setting off what for centuries of future critics was to become the most infamous and mysterious literary rivalry in all of Spanish literature. Just who was the Avellaneda who chose to hide behind an obvious pseudonym? Why did he decide to write a spurious sequel and rush it into print a year before Cervantes could complete his own continuation? More importantly (at least since the beginning of the last century), what was it about the imitator's treatment of Don Quixote that was so disturbing and so unlike the original? Why, in other words, does it strike the modern reader as violent?[1]

The riddle of Avellaneda's true identity has never been solved nor have the origins of his personal dislike for Cervantes been entirely recovered.[2] As for the novel itself, it was received with indifference by the Spanish reading public and would only gain new life in a French adaptation published ninety years later. By the nineteenth century, the canonization of Cervantes' two texts had decided the fate of the "false" *Don Quixote*, consigning it to the realm of cultural oddities which were deemed both unreadable and undesirable. The banishing of the other Quixote not only served to dehistoricize the Cervantine text, creating the impression that Cervantes himself was the anomalous "inventor of the modern novel" (hence his status as virtually the only Spanish author to be allowed into the world-literature canon), but also contributed to an on-going process by which our understanding of power and subjectivity in the early modern period has been constructed. What I want to argue in this essay, then, is that the uses to which the figures of Don Quixote has been put since its literary inception have been inextricably bound up with the process of political contestation and legitimation not only within the original Spanish context but in subsequent historical moments as well. Until we allow the other version, Avellaneda's Quixote, to stand in the open next to that of Cervantes, we will neither see the complexity of the cultural formations which struggled to control meaning in seventeenth-century Spain nor understand the way in which the two Quixotes finally became one. Put another way, we must begin the process of unmasking these two distinct yet interrelated forms of violence (Avellaneda's representation of violence and the violence with which Cervantes' text was removed from the conflictive scene of its inception) which, left concealed, works to prevent literature from being viewed as the site of struggle and contradiction.

Crucial to our understanding of this contention is the agonistic dialogue which was often set up between individual texts, an intertexuality in the strong sense having less do with strictly literary resonances than with specific political and ideological oppositions. The rivalry between Mateo Alemán, for example, and the author of a spurious continuation of his *Guzmán de Alfarache* (1599) was a contemporary antecedent of the Cervantes–Avellaneda dispute. Still another case of this kind of competition in early modern Spain was the one involving Francisco de Quevedo's picaresque novel *El Buscón* (1626, but probably written *c.* 1604) and the anonymous *Lazarillo de Tormes* (1554). In essence, Quevedo's project is to undercut the social criticism and suggestion of mobility implicit in the earlier text by punishing a similar kind of protagonist (the *picaro*) who fights to better his status within the limited range of possibilities available. This is accomplished not only by direct physical violence directed against the hero, but also by a stylistic transformation that reifies everyday reality and aspires to a kind of destruction of the referent extreme even for the Baroque period. The battle that will be waged between the two versions of Don Quixote shares many of the strategies that constitute the picaresque rivalries, yet the stakes are much higher since the Cervantine project is ultimately more challenging to the established order than is that of the unknown author of the *Lazarillo*. To map the ideological terrain of seventeenth-century Spain exhaustively is beyond the scope of this essay. Nevertheless, there are clear signposts that help us to see exactly where the principal antagonisms reside.

One of the guiding axioms (if not the central principle) of Cervantes' text is the concept of "Cada uno es hijo de sus obras" ("Each man is the child of his actions"). This idea, as it is articulated within the double project of Arms and Letters, forms the basis of Don Quixote's exploration of subjectivity and, despite the protagonist's status as a minor nobleman (*hidalgo* – "child of something"), is developed in direct opposition to inherited nobility, the third element considered to be central to the determination of identity within aristocratic discourse. Thus in the Cervantine text, with the exception of an occasional remark by Sancho Panza followed by Don Quixote's rebuke, there are virtually no references to inherited nobility. As we listen to Avellaneda's Quixote, Cervantes' conscious omission is clearly foregrounded:

> Indeed, gentlemen, I think it would be hard to find three people like us to travel from Zaragoza to this spot, for each of us is deserving of honor and fame because, as we know, one of three things in this world assures us of them: blood, arms, or letters, virtue being common to all three, making a perfect combination. (Avellaneda 1980: 215–16).[3]

Cervantes' denial of blood as a way in which one is recognized in the world sets him radically apart from the dominant axiological system of his society. Carried too far, the suggestion that inherited social position or lineage had little to do with a person's value could conceivably have brought down the entire cultural apparatus which sustained Spanish absolutism: aristocratic privilege, the exclusion of impure ethnic groups (Jews and Muslims), the ability of the community to employ shame as a controlling device, the "cleansing" of the blood through revenge (a ritual act crucial to an understanding of Calderonian theater) and so on.

What we see in Avellaneda's version of Don Quixote, I would argue, is the disciplining of Cervantes' character (and the kind of subjectivity it figures forth) by means of two complementary modes of violence: shame and revenge. Although I will stress the importance of shaming strategies in the novel, this is not to say that direct corporal punishment no longer formed the basis of the law in seventeenth-century Spain; on the contrary, the techniques made infamous by the Inquisition ever since its founding in 1480 were only the most dramatic examples of a well-oiled machinery of discipline which functioned at all levels of Spanish society during the period we are discussing. Yet in the Spanish context the experience of violence designed to reposition the wrongdoer in society (hence necessary to the maintenance of order) was almost always informed by a powerful ascetic inheritance which targeted the individual consciousness as the domain of control to an equal if not greater extent than the body. Foucault has spoken in passing of shame as part of the economy of punishment, yet the seductive descriptions of public torture in *Discipline and Punish* have tended to make us forget the functioning of this rather less explicit manipulation of the subject. For Avellaneda, it will not be necessary to make Don Quixote suffer any pain beyond that produced by the humiliating situations in which he is placed since his protagonist is little more than a lunatic who is able to produce his own spectacles.[4]

II

> Shaming another in public is like shedding blood. (*Talmud*)

It is no accident that the Avellaneda Quixote's first act as a "reactivated" knight-errant is to interfere with a public punishment.[5] Newly-arrived in the city of Zaragoza, he is almost immediately confronted with the reality of power as it is practiced by the absolutist state:

> It so happened that as Don Quixote was going up the street, giving everybody who saw him looking as he did plenty to talk about, the law was bringing in a man riding on a donkey. His back was naked from the waist up, there was a rope around his neck, and he was being given two hundred lashes for being a thief. Three or four constables and notaries were accompanying him, with more than two hundred boys following them. (Avellaneda 1980: 73).

Don Quixote attempts to free the thief since in his madness he believes him to be a knight who has been unjustly accused, yet soon finds himself overwhelmed by the officers of the law: "they took him off Rocinante and, to his grief, tied his hands behind his back; gripped firmly by five or six constables he was dragged off to jail. . . . They put his feet in the stocks and handcuffed him after having taken away all his armor" (ibid.: 76).

Don Quixote's lawlessness is quickly controlled, then, and he is rapidly subjected to the same position as that of the thief he had tried to liberate. This of course is only one of the many occasions Avellaneda's Quixote will find himself in chains, a condition not experienced by Cervantes' hero in either the 1605 or 1615 novel. Physical imprisonment in the original text (with the exception of the open cage in the final chapters) is displaced by a series of images designed to explore the inescapable yet complex and often tenuous interrelationships between the self and the social world (ropes, strings, nets, etc.).[6] In Avellaneda, on the other hand, the subject must be held forcibly in place so that it can be the recipient of both the law's violence and the shame produced by exchanges with the community ("all the people at the jail door were saying, 'The poor knight in armor well deserves the lashes that await him, because he was so stupid that he meddled with the law . . .'" (ibid.: 78)). The condemnation of Don Quixote by the subordinate classes merely underscores the ideological homogeneity to which the Spanish elites (with Avellaneda as their spokesman) aspired. In this text (unlike that of Cervantes) there will be none of the willing participation in the quixotic project by innkeepers, serving girls, or galley slaves.

Yet, despite his frequent bondage, the other Don Quixote will not be disciplined with an orchestrated attack on the body. Despite his inciting a spontaneous beating by the jailer, vividly described for us by the narrator (''half a dozen punches in the face, making blood stream from his nose and mouth'' (ibid.: 76)), Avellaneda's Quixote will be rescued from the hands of the civil authorities (''el Justicia'') by the aristocrat (Don Alvaro) who will later trick him into the asylum in order to undergo a different kind of punishment whose efficacy originates in public humiliation and sexuality itself. The reader's experience of this process begins soon after Don Quixote's rescue. Left alone by Don Alvaro in order to rest, he imagines that he is addressing the judges at a jousting tournament:

> A la voz grande que dio, subieron un paje y Sancho Panza; y entrando dentro del aposento, hallaron a don Quijote, las bragas caídas, hablando con los jueces, mirando al techo; y como la camisa era un poco corta por delante, no dejaba de descubrir alguna fealdad; lo cual visto por Sancho Panza, le dijo:
>
> -Cubra, señor Desamorado, ipecador de míl, el etcétera, que aquí no hay jueces que le pretendan echar otra vez preso, ni dar docientos azotes, ni sacar a la vergüenza, aunque harto saca vuesa merced a ella las suyas sin para qué; que bien puede estar seguro.
>
> Volvió la cabeza don Quijote, y alzando las bragas de espaldas para ponérselas, bajóse un poco y descubrió de la trasera lo que de la delantera había descubierto, y algo más asqueroso. Sancho, que lo vio, le dijo:
>
> -¡Pesia a mi sayo! Señor, ¿qué hace? Que peor está que estaba: eso es querer salundarnos con todas las inmundicias que Dios le ha dado. (Avellaneda 1971: 153)
>
> (At this loud shouting a page and Sancho Panza came upstairs; entering the room they found Don Quixote with his breeches down, talking to the judges and looking at the ceiling. Since his shirt was a little short in front, it didn't fail to reveal a bit of ugliness. When Sancho Panza saw this he said, ''Cover your etcetera, loveless sir! Sinner that I am! There are no judges here intending to take you prisoner again or give you two hundred lashes or take you out to be shamed, even though you are only too well uncovering your shame with no reason, you can be sure!''
>
> Don Quixote looked around, and as he lifted the breeches to put them on with his back turned he bent over slightly and revealed from the rear what he had revealed from the front as well as something more repulsive. Sancho, who saw it, told him, ''Confounded be my breeches, sir! It's worse than it was before; what you are doing is trying to salute us

> with all the unmentionable things God has given you'' (Avellaneda 1980: 85)

Besides the explicit humiliation of the protagonist and the ascetic abhorrence of the body, this passage reveals one other detail important to my argument. Sancho's pun on *vergüenza* is crucial here because it reminds us that within the semantic field of seventeenth-century Spanish this word signified public punishment, shame, and the genitals.[7] Covarrubias' *Tesoro de la Lengua Castellana o Española* (1611) clarifies the connection between sexuality, shame, and punishment:

> VERGÜENÇAS y partes vergonçosas en el hombre y la mujer, *latine pudenda, a pudore*. Sacar a uno a la vergüença, es pena y castigo que se suele dar por algunos delitos, y a estos tales los suelen tener atados en el rollo por algún espacio de tiempo, con que quedan avergonçados y afrentados.
>
> (VERGÜENÇAS and shameful parts in men and women, *latine pudenda a pudore*. To take someone out for public shaming is a sentence and punishment which is usually given for some crimes, and these types of offenders are usually placed in the pillory for some time so that they are ashamed and affronted.)

The word *vergüenza* and its multiple significations is strikingly present throughout the entire section beginning soon after Don Quixote's arrest (''they were already starting to unchain him from the stocks in order to take him out for public punishment [*vergüenza*],'' to his rescue by the nobleman (''Don Alvaro said, 'They would undoubtedly have put you to public shame [*vergüenza*] if your good fortune, or better said, if God, who settles all things smoothly, had not arranged my arrival' ''), to the episode with Sancho I have quoted at length. For our purposes and with regard to the determination of subjectivity, it seems clear that within the economy of punishment shame was an equally if not more important constituent than direct physical pain.[8] The psychological impact on the ''criminal'' or deviant was perhaps less crucial than the overall ''scene'' itself in which the community (in the case of a literary text, a community of readers) marked the person to be shamed as deserving of ridicule and reasserted its values over and against individualized behavior. Thus it would seem that this kind of spectacle has less to do with the display of the ''unrestrained presence of the sovereign'' (as Foucault would have it) than with the complicity and participation of the entire social body in the containment of anomalous subjectivity.[9]

The degradation suffered by Avellaneda's Quixote in the episode I have just discussed takes an additional turn moments later. Having been asked to dine with Don Alvaro, Don Quixote is visited by a small group of nobles:

> They had been invited in order to decide on the livery each was to wear and to enjoy Don Quixote as though he were a private performance, so

> they went straight up to his room and, finding him half-dressed and looking as has been told, they laughed heartily. . . . He [Don Quixote] used such extraordinary names that at each one the guests gave a thousand retches of suppressed laughter. (Avellaneda 1980: 87)

The shifting of location for his new humiliation replaces the peasant spectators of the first scene of shame (Sancho and the page) with those drawn from a more powerful social group, not so much to correct Don Quixote's behavior as to make each member of the aristocratic audience (both inside and outside the text) secure in his own sense of superiority: the grotesque body of the marginalized individual is forcefully ridiculed. On this view, for those who "rescue" and subsequently mock him, Don Quixote's mental infirmity is far less important than his social status – an *hidalgo*, the lowest rank within the minor nobility – and the belief that it is the aristocracy's privilege (not the state's) to discipline a member of their own social group who acts outside the range of accepted behavior. All of this has little to do with a re-enactment of feudal relations (the humbling of a peasant by an aristocrat), yet we must remember that Don Quixote is not a character drawn from the subordinate classes nor did feudalism on the classic model ever develop in Castile; its representation is relatively rare in Spanish literature even in medieval epic texts such as the *Poem of the Cid*. Don Quixote's exposure before Don Alvaro and his colleagues, then, is not a sign of the struggle between estates, but rather is an example of the conflict between sub-groups within the same social formation – in this case, the wealthy urban nobility's "correction" of an impoverished and unruly *hidalgo* from the countryside.

As we have seen in the two scenes discussed above, the economy of violence in Avellaneda's text consists primarily of shaming techniques designed to reposition the subject in terms of the established social order. When the offense is sufficiently grave, however, shame will be supplemented by personal revenge and a direct physical attack on the offender. That Spanish literature of this period was virtually obsessed by the theme of *personal* vengeance is a symptom both of the incompleteness of Spanish absolutism as well as Spain's continuing implication in earlier cultural systems (Mediterranean honor/revenge societies). It is also clear that the state's growing monopoly on revenge could not be easily figured within literary discourse and had to be displaced back on to earlier cultural models. Thus in the first interpolated tale which Avellaneda situates at the center of his text (a structural imitation of Cervantes' 1605 novel), the reader is shown how the transgression of social norms can lead not only to public disgrace, but in extreme cases to physical and moral destruction.

Briefly told, the story concerns a Flemish aristocrat, M. de Japelin, who is inspired by a Dominican friar to enter the priesthood. After doing so, Japelin is suddenly overcome by doubt and decides to leave the order and marry. The birth of their first child is soon followed by the couple's receiving as a house guest a Spanish solider who immediately falls in love with the lady (herself a former novice) and through a series of deceptions and mistaken identities succeeds in having sex with her. Overcome with shame and fear, he flees the house only to be pursued and murdered by an enraged Japelin. His wife having already committed suicide, Japelin returns home to kill his newborn son and himself.

What is striking about this otherwise typical exemplary tale is the way in which all the main characters are violently punished. The Spanish soldier who forces himself upon the unknowing wife (only days after her having given birth to a son) is depicted as regretting his act: "returned to his room and bed, quite grief-stricken over what he had done. Since guilt is followed by repentance and sin by shame and worry, he was at once so overwhelmed by his evil that he cursed his lack of reasoning and patience" (Avellaneda 1980: 44). Yet he nevertheless becomes the object of Japelin's merciless vengeance. After running the Spaniard through with his lance, "he dismounted, pulled the javelin out of the corpse and stabbed it again five or six times, hacking its head to pieces. . . . The soldier remained there, trampled in his own blood, food for birds and beasts" (ibid.: 149). The murder of the infant is no less graphic, nor is Japelin's own suicide which ends the chain of violent episodes.

Any reader familiar with Cervantes' *Don Quixote* will recognize such passages as being utterly foreign to the original novel. And, as I have been arguing, Avellaneda's project is significantly different from that of his precursor. By constructing his text upon the twin constraints of shame and revenge, Avellaneda drives home his point that the attempt to speak another kind of subjectivity through the abandonment of one's social position can only lead to death and damnation. As Avellaneda's priest says upon hearing the story of Japelin: "The sad end of all the main characters in that tragedy is very much to be feared, but the principal actors could expect nothing better (morally speaking), since they had given up the religious orders which they had started to join" (ibid.: 153).[10] The exploration of alternate and more flexible social relations is impermissible in a culture founded on the maintenance of aristocratic and clerical privilege, and the application of force is always an option should internalized controls fail.

It will be argued that these are purely fictional scenes of violence and punishment, having more to do with past literature than with the reality of daily life in Spain. In so far as Japelin's revenge is personal, we may say that it was the exception to the norm, yet accounts of similar attempts to test the *status quo* are numerous throughout the seventeenth century. In a letter written in 1635, for example, a Jesuit reports that in the village of Piedrahita an elderly nun, having fallen in love with an elderly gentleman, had abandoned the convent in order to run off with him. Fleeing towards Portugal, they reached Ciudad Rodrigo: "There the police of Avila (under whose jurisdiction Piedrahita is) came up with them and took them; she returned to her convent where she will die imprisoned or immured; he was taken to Avila, where today he was executed.''[11] In this case, then, it is the state that is responsible for and no less expeditious in carrying out the elimination of dislocated subjects. Avellaneda, however, in his attempt to contain the Cervantine experiment, allows Don Quixote to live, albeit with significant preconditions. If it is personal vengeance that supplements shame in the example of Japelin, and the direct force of the law which ends the lives of the wayward nun and her companion, it is confinement that must mark the final resting place of the other Quixote. The series of public humiliations which seek to discipline him in lieu of direct physical punishment culminates in the final scene of the novel set in the Toledo madhouse, a space in which the chorus of dissonant voices can be more easily subjected to both indoctrination and readjustment.

III

> No one can be successful in the company of the mad unless he converts to insanity before. (Judah Ha-Levi)

In *Madness and Civilization*, Foucault tells the story of Renaissance madness floundering about in broad daylight. Not yet confined within the fortress of reason, unreason was free to effortlessly move back and forth across the still unguarded border of the public sphere. Before the shift in epistemes that would inaugurate the new age, Don Quixote traversed a landscape as yet unmarked by institutional enclosures designed to control those who were thought to be insane. Thus runs Foucault's account. Its strong sense of wandering and rurality reminds us of Foucault's ''linguistic'' Don Quixote who, as he is elaborated in *The Order of Things*, inhabits a second frontier between words and the things they represent.

It now seems clear, however, with the other Quixote out in the open, that by 1614 there were already strategies for the containment of madness, well-established strategies which revealed a strong desire on the part of some sectors of society to remove the ''madman'' from public view in order to inaugurate a cure. Foucault's account of Renaissance madness, I would argue, privileges the humanist tradition represented by Erasmus, Cervantes, and Shakespeare in order to make more dramatic the moment of the ''great confinement'' of the later seventeenth century. What is given insufficient attention by Foucault is the counter-tradition of ascetism which, in its open hostility to the mad as monstrous and inhuman, continued to thrive throughout the early modern period and in Spain culminated in the satirical writings of Francisco de Quevedo and others. On the level of Spanish institutions, this other view of madness contributed to the early dismantling of the Ship of Fools, and its rotting timbers were rapidly converted into the rafters of the urban Hospital. Indeed, the first asylum in western Europe devoted specifically to the mentally ill was the one founded in Valencia in 1409 (Lope de Vega set his play *Los locos de Valencia* here), and throughout the fifteenth century mental institutions were established in most of the major cities of Spain: Zaragoza (1425), Seville (1436), and Valladolid (1489). Avellaneda's Quixote, then, while not subjected to the bonds of classical reason which would be forged in the eighteenth century, nor retrained in order to increase his economic productivity, is nonetheless "made docile" by recognized forms of domination underwritten by the social powers of the age: the aristocracy and the Church. It is no accident that in Avellaneda's novel Don Quixote is led by Don Alvaro Tarfe, a nobleman from Granada, to the Casa del Nuncio, the Toledo asylum founded in 1483 by a representative of the Holy See and infamous for having the worst conditions of any asylum in Spain. As Don Quixote waits patiently in the courtyard, unaware that he is about to be committed, one of the inmates tells him: "these thieving guardians are bringing you to throw a heavy chain on you and give you a thorough trouncing until you recover some sanity whether you like it or not. They've done the same thing to me" (Avellaneda 1980: 339). Despite the obvious illness of the other Quixote, the clarity of the inmate's warning suggests that the issue of sanity has become secondary to broader sociocultural concerns. Avellaneda imprisons his protagonist because he is mad, yet his more pressing ambition is to castigate the alternative forms of subjectivity which Cervantes had explored in his novel. The attempt to cultivate an individualized existence runs head on not only into the traditional constraints of shame and revenge, but finally into the twin powers of early Spanish absolutism (with Avellaneda as their literary spokesman) which work to suppress that attempt.

The transporting of madness into an urban environment, then, converted its harmless meanderings into a new economy of violence, punishment, and confinement, and became part of the larger design aimed at controlling all radically individual behavior. It should not surprise us that Avellaneda's novel is set primarily in cities while Cervantes sets his in the country. As one of the characters in Lope de Vega's *Fuente Ovejuna* says: "En las ciudades hay Dios/y más presto quien castiga" ("In the cities there is God's justice/and rapid [secular] punishment") (1985: II.1007–8). It is not until the 1615 *Quixote* that Cervantes would take his protagonist into the courtly society of the duke and duchess and the properly urban setting of Barcelona. In the prologue and first chapter of his second novel Cervantes prepares the reader for the new landscape in which Don Quixote will move, giving the reader three accounts of madness, two in the streets of Seville and Cordoba, one in the asylum at Seville. Yet at no time is Cervantes' Quixote exposed to the horrors of the madhouse, and even though the literary experience of the asylum in Cervantes reveals the same relations of power as we see at work at the end of Avellaneda's version (Church and state), there is none of the physical violence which marks the latter's description of the protagonist being attacked by one of the inmates ("a sudden rage came over the lunatic and he bit him viciously two or three times, ending by seizing his thumb between his teeth in such a way that he almost lopped it off," (Avellaneda 1980: 343–4). Cervantes, opposing his text to that of his imitator, presents us with a madman who, having convinced the archbishop of his sanity through a series of letters, in essence (since in reality he continues to be mad) mocks the traditional knowledge which had been offered to Avellaneda's Quixote as he was being committed to the Toledo madhouse.[12] The undecidable status of the graduate who thought he was Neptune and Don Quixote's response to the barber's story underscores Cervantes' belief that his character's sanity or madness is not a condition to be decided upon by external sources, but rather is an act of will inextricably linked to Don Quixote's exploration of subjectivity.

In the end, the pressure of traditional values alienates both Don Quixotes, yet only Cervantes' hero has the self-consciousness to realize his intolerable position *vis-à-vis* the elites who seek to manipulate him. Avellaneda's Quixote in fact displays a strong desire to ingratiate himself with the powers-that-be, and early in the novel we are told that his ultimate goal is to be defined within the confines of aristocratic society: "he intended to go to the court of the King of Spain to make himself known by his exploits. 'And,' added good Don Quixote, 'I shall make friends with the grandees, dukes, marquises, and counts who help in the service of his royal person'" (Avellaneda 1980: 30). Despite the fact that he intends to provoke and kill the majority of the noblemen he meets, Avellaneda's Quixote is determined to surrender his subjectivity to the economy of courtly society ("His Catholic Majesty will be forced to extol me as one of the best knights in Europe") and thereby be incorporated into the body politic. All of this stands in sharp contrast to Cervantes' 1605 character who virtually never speaks of the Spanish ruling classes and at no time expresses the wish to gain their favor or serve their interests. The refusal to align himself with any of the hegemonic forces at work in his society underscores the strong sense of wandering which has been noted by critics from Lukács to Foucault. By moving along the margins of the dominant culture, Cervantes' Don Quixote is not the romantic rebel which much of traditional criticism has wanted him to be, yet he nonetheless works to clear a space upon which modern forms of subjectivity would later appear.[13]

Some seventy-five years before the first appearance of a literary Quixote, the Spanish elites had effectively repressed any further development of Erasmian thought on the Iberian peninsula, although material taken from the original humanist program continued to resonate in later literature. If the hero of Cervantes' 1605 novel had not posed a threat (even if only on the symbolic level) to the order which surrounded him, there would have been little need for Avellaneda's violent response. By no means was Cervantes proposing a revolutionary concept of subjectivity (nor even a fully-developed bourgeois concept); this simply was not available to him in the historical juncture in which he was situated, and as I have already mentioned the 1615 novel contains several changes in the Quixotic project similar to those Avellaneda had made: the suggestion that knights errant might serve the king (II.1), the introduction of Sampson Carrasco as the agent of Don Quixote's "cure" (II.3), the subtle display of state-sponsored revenge (II.60), and the spectacle of public humiliation in Barcelona (II.62). In addition, the novel's final scenes emphasize the fact that Cervantes was content to have Don Quixote die a traditional death well within the bosom of the Church. Yet Avellaneda could not have foreseen these developments in the later novel. Sufficiently disturbed by the gesture towards a new kind of subject found in Cervantes' first text, he responded quickly to punish and correct it. What this literary rivalry brings into view, then, is a struggle that would be carried out on diverse cultural battlegrounds for decades to come. In Spain, the emergent form of subjectivity sketched out in Cervantes' 1605 text would be locked away in the gloomy dungeons of Avellaneda's madhouse at least until the early nineteenth century. In other parts of Europe, the Cervantine Quixote would ride at the vanguard of bourgeois individualism and the new forms of freedom and containment it authorized.[14]

IV

> Why, Don Juan, do you want to read this nonsense? Can anyone who has read the first part of the history of Don Quixote de la Mancha possibly take any pleasure in reading the second? (Don Jerónimo, a character in Cervantes' 1615 novel, discussing Avellaneda's *Don Quixote*)

If, as recent criticism has argued, violence is central to the maintenance of all social formations then we might well conclude that Cervantes' text merely signals a new kind of violence no less brutal than the one represented by Avellaneda. While I would agree that in essence all discourse involves the use of power, I would also argue that we need to distinguish between different kinds of power at work within the historical conjuncture we are studying. In calling for a normative analysis of the two Quixotes, I assume that all social practices are not equivalent but rather that at certain moments some may serve the "radical" function of shifting the conventional gridlock which overlays daily life, thereby rearranging the sites at which both political and representational power are produced. On this view, given the dissimilarities between Avellaneda and Cervantes we have seen at specific points in their novels, it is clear that Cervantes' model of subjectivity, by stretching the traditional limits set by a predominantly aristocratic and religious culture, clears new territory upon which a wider range of subject positions can be realized. To implicate Cervantine humanism in the institutional violence produced by twentieth-century liberal discourse would be a kind of presentism which replaces transhistorical values and content with the no less totalizing category of "neutral" power.

Having said this, I would not want to be misunderstood as attempting to defend the canon at any cost. That a later variation on the bourgeois subject towards which Cervantes' Quixote only haltingly signals will seek to exclude other heterogeneous elements (women, ethnic minorities, etc.) cannot be denied. But it was not until the long process of the consolidation of the bourgeoisie had been completed that Don Quixote took his place at the front line of this struggle. It should not surprise us that England was the first country to adopt Cervantes' character since, as I have argued elsewhere, the proto-modern concept of the self would enjoy a brighter, albeit precarious, future in England than it would in Spain.[15] The Shelton translation of 1612, once united with an English version of the second novel (1620, also by Shelton), began a slow process which would result in the apotheosis of *Don Quixote* as the literary Bible of idealist aesthetics and bourgeois individualism, and assure that its complementary enemy (Avellaneda) be consigned to anthologies of literary curios. We cannot forget the role played by Goethe, Hegel, Schlegel, and German Romanticism generally in the legitimation of Cervantes' text, yet it was in eighteenth-century England that *Don Quixote* began to enjoy the status it continues to hold today within the humanistic canon. Until as late as 1703 in England, Cervantes was considered to be little more than a "comick author," a label which had been attached to his name in Spain as well ever since the publication of his 1605 novel. It is in this historical moment, however, at the beginning of the Enlightenment that the paths of Cervantes and Avellaneda were destined to cross once again, an event which contributed to a critical re-evaluation of both authors.

In 1704, the French novelist and critic René Lesage, author of *Gil Blas*, translated Avellaneda's spurious sequel to *Don Quixote*. Lesage's text is more an adaptation, full of omissions, additions, and changes, than a direct rendering of the original, but it was this text that would have an immediate impact on the future of both Avellaneda and Cervantes since its elaboration of Avellaneda's central project (the "correction" of Cervantes' 1605 text) served as a screen through which the real *Don Quixote* could now be reread. Lesage's especially strong dislike of the first Cervantine novel (an opinion which would not be shared in the nineteenth century by Goethe, Coleridge, or others) and overall preference for Avellaneda's version is made a topic in the French edition, and in fact Lesage accuses Cervantes of having plagiarized much of his 1615 continuation.[16]

It is in Lesage's revised ending, however, that we see most clearly the way in which Cervantes and his character were punished by the social powers of eighteenth century French society. Outside his village of Argamasilla, Don Quixote confronts a group of soldiers from the Holy Brotherhood (a rural police force), attacks them verbally, and insults them as knights in the service of an unworthy prince:

> The officer, who commanded the party, thinking these words reflected on the king his master, answered Don Quixote thus. Sure thou art mad, or some damned insolent fellow, that darest speak such words of the most honourable of all princes. Don Quixote hearing himself called madman, and damned fellow, set himself fast in his saddle, couched his lance, and ran full tilt at the officer, who having neither time nor skill to put by, or avoid the thrust, received it in his heart, and fell down stone dead under his horse's belly. Then the troopers drew their swords and hemmed in the knight to seize him; but he drew as well as they, and charged them so furiously that he wounded two or three of them. The others fearing the same fate, began to give way, when one of their companions ashamed that the whole party could not secure a single man, laid hold of his carabine [*sic*], and taking aim at Don Quixote's face, shot him through the head with a brace of bullets. The poor knight had no need of a second shot. His feeble hand dropped Rocinante's bridle, and after tottering a while in the saddle, he at last fell off, near the dead body of the officer he had slain . . . .
>
> Whilst Sancho thus raved, the curate Peter Perez and the barber came upon the field of battle, and finding no signs of life in Don Quixote, were much troubled. The troopers would have taken possession of Don Quixote's body to form a process against him as a common disturber of the peace, to render him and his memory infamous[17]

I have quoted Lesage at length because his text reveals the ways in which an aristocratic reaction to the Cervantine experiment within a new context could be even more severe than Avellaneda's. The extreme violence initiated by Lesage's Don Quixote (Cervantes' character wounds several people but kills no one), and brutally answered by the protectors of social order, is symptomatic of the rigid controls placed on the subject in the early decades of the Age of Reason. Here, it is not enough that madness be removed from the scene of culture as it had been both in Cervantes (through a self-conscious act) and in Avellaneda (through incarceration). The summary execution of the unruly subject must be followed by the inscription of the lifeless body into the discourse of the law in order to insure that its marginalization continue well beyond the present historical moment.

The French version of Avellaneda was quickly translated into English (1705 and again in 1707), and in *An Essay on Criticism* (1711) Alexander Pope refers to Don Quixote and cites an episode invented by Lesage despite the fact that in his pamphlet "The Critical Specimen," Pope calls Avellaneda "a false Cervantes." In the *Dunciad* (1728) Pope moves back to the original author and speaks of "Cervantes's serious air," a fact which shows that the English poet had read both of the variations on Don Quixote and considered at least the original to be something more than a mere jest-book. These references are crucial since they show that Cervantes' text, now considered legitimate as serious literature, had not yet achieved that status through the definitive exclusion of the imitator's text; on the contrary, the two novels continued to be read side by side and in fact one was often confused with the other. Throughout the century, then, these two versions, not yet perceived as being antithetical, would be reworked by the great novelists of the middle class. Sterne, to cite only one example, proclaimed Cervantes as his favorite novelist, but at the same time did not hesitate to adapt certain passages of Avellaneda–Lesage.[18] By the turn of the century, however, the slow process by which Cervantes' text would be placed above those of its imitators once and for all was already under way.

It was in 1805 that English readers finally possessed an authentic edition of Avellaneda, although in a sense it was too late for this text which had shadowed the real Don Quixote ever since 1614. The reformulation of literary discourse by both idealist philosophy and bourgeois social relations made it inevitable that Cervantes would be extricated from the struggle with his competitors in order to be sanctified as a classic author. By the 1880s, Cervantes' popularity had become so great that three different editions of his text were published in English within a period of seven years (1881, 1885, and 1888), and soon after the first annotated edition in any language was also published in English. The influential literary pundits of the day raised *Don Quixote* to the level of a religion: "Macaulay esteemed it 'the best novel in the world, beyond all comparison' . . . while to FitzGerald it became 'the Book.'"[19] In this new historical conjuncture, what was perceived as the "rebellious" subjectivity of the 1605 *Quixote* towered over the aristocratic and traditionalist ideologies which informed the texts of both Avellaneda and Lesage and made them undesirable. Avellaneda's attempt to punish had finally failed, at least in the English-speaking world. What had always been the two Quixotes had now become one.

The reason we read only the "true" *Don Quixote* today, I suggest, is because the rewriting of its tentative insights into emergent forms of subjectivity validates our own sense of self, of literature, and of the humanities in general. What has been lost, what the separation of violence and literature has done, is to silence the other *Quixote* and in the process mask the ideological struggle in which Cervantes' text was engaged, not only with that of Avellaneda but with the vast majority of seventeenth-century cultural production in Spain. Keeping in mind the important differences between the two great imperial and literary powers in the early modern period, I would argue that what Leonard Tennenhouse has written of Shakespeare might also be said of Cervantes: "we must assume his art was always political and that it is our modern situation and not his world of meaning which prevents our finding his politics on the surface and seeing his strategies of displacement as political strategies."[20] And we must surely hasten to add: a political Cervantes is no less great a writer than an aestheticized Cervantes.

The depoliticizing of *Don Quixote* created a transcendental work of world literature at the price of mystifying its original context as well as our own. This is to say that the appropriation of *Don Quixote* I have briefly outlined above constituted a more subtle form of power which continues to conceal the ways in which discursive practices delimit potential forms of subjectivity. Liberal humanism's central myths of individualism, creativity, and freedom (with Cervantes' hero as critical example) blind us to its own strategies of confinement and exclusion which are only slightly less violent than Avellaneda's madhouse or Lesage's shotgun. The other *Quixote* can no longer be read not because it sought to repress an incipient version of the modern subject, but because it reminds us that all forms of subjectivity are conceived in a bitter struggle for power and hegemony.

Notes

1 The way in which I will speak about violence has to do primarily with the practices of shame, revenge, and physical confinement as displayed in novelistic texts and finally with matters of canon formation, appropriation, and cultural hegemony. That these latter issues constitute a kind of violence will be a basic premise of my essay; the limits placed on subjectivity by liberal discourse are no less real than the constraints we will see at work in Avellaneda's text. Yet as a Hispanist I am immediately struck by the fact that any discussion of violence as rhetoric, representation, or discourse necessarily backgrounds the everyday reality of brute force faced by millions of people in places where this volume will never be seen (such as Central America). Just what the relationship might be between these more pressing struggles and a collection of theoretical essays is a complex problem which I hope we as members of the academic community will not fail to address.

2 My sense of the problem of Avellaneda's true identity is that it is relatively unimportant in terms of how the text functions within the culture it was produced. It seems likely, however, that the author was a Dominican friar, a fact that should surprise no one given that the novel speaks throughout from a dominant ideological position. On these and other matters, the superior study of Avellaneda continues to be Gilman (1951). One other critic who has understood Avellaneda's text as a further problematizing of Cervantes' exploration of identity is Riley (1962).

3 Citations in Spanish are to Avellaneda (1971); English versions are from the fine translation (the first in 200 years) of Avellaneda (1980). English quotations of Cervantes are from Cervantes (1982).

4 It will be argued that since Avellaneda's Don Quixote is completely mad (unlike Cervantes' character who is also known for his "lucid intervals"), he is incapable of feeling shame or even pain. My contention throughout this essay, however, will be that on every occasion that Avellaneda punishes his protagonist, he is indirectly punishing Cervantes' character as well and by extension repudiating the forms of subjectivity represented in the 1605 text.

5 The resonances of Cervantes' own episode of Don Quixote and the galley slaves (I.22) are strong here, yet there is not the same discussion of liberty nor the final unhindered dispersal of all the participants. Ginés and his colleagues do not punish Don Quixote, but rather attack him for attempting to limit their newly-found freedom ("it is my will that you bear this chain which I have taken from your necks and immediately take the road to the city of El Toboso, there to present yourselves before the Lady Dulcinea del Toboso"), and then flee.

6 For a discussion of this kind of imagery as a narrative device, see Ponsetti (1981).
7 The close connection in early modern Spanish between shame and public exposure is especially interesting since the etymology of English "shame" links it more closely to covering up and hiding. The Spanish usage would seem to support my contention that there was as yet no strict division between the public and private spheres in Cervantes' Spain.
8 An essential part of inquisitorial proceedings was the undressing of the accused prior to actual torture. See Beinart (1974).
9 Paradoxically, the recuperation of the aberrant self into the cultural entity through shaming techniques would seem to contribute simultaneously to the individuating process by at least a momentary setting apart of the guilty party from the community. Foucault in fact has spoken of a process of inversion by which the individual being punished, having been temporarily detached from the social body, is converted into a hero. These matters, however, are more properly suited to the fields of social psychology and anthropology, and will not be pursued here. I should add that I owe the epigraph to section II of this essay to Léon Wurmser (Wurmser 1981).
10 The priest's remarks, with the explicit warning about leaving the priesthood, as well as the content of the second interpolated tale, seem to support the hypothesis that Avellaneda's audience was initially a clerical one and that he himself was a Dominican. Although Avellaneda's second interpolated tale is less concerned with the issues central to my argument, it also exemplifies the extent to which elements of an earlier theological code are reworked for a seventeenth-century audience. The story of a prioress who leaves the convent to pursue a life of secular debauchery, is abandoned, and finally forced to prostitute herself, is drawn from the miraculous discourse of Mariolatry more common to the fourteenth century in Spain – after returning to her convent, the nun discovers that the Virgin has protected her good name by playing the role of prioress in her absence. As far as the congregation is concerned, the nun had never been gone. The restoration of order disrupted by the abandonment of social position and holy orders, which in the first tale can only be accomplished through violence, is here achieved by the direct intervention of the sacred.
11 Letter of Father Juan Chacón dated December 2 1635. Quoted in Baroja 1966: 113. Four years after Father Chacón's letter, Sebastián Coto would devote an entire text to such problems (Coto 1639).
12 The page of the nobleman who had arranged for the knight's confinement (known to Don Quixote as the Archipámpano) tells Don Quixote: "Take care of your soul and realize God's mercy in not permitting you to die on those roads in the disastrous situations in which your madness placed you so many times" Avellaneda 1980: 344). In the barber's story of Cervantes' 1615 text (II.1), the madman of Seville decides that "by the mercy of God he had recovered his lost wits," a fact refuted a few pages later when he proves that he has only played at being sane and is led back to his cell. In Cervantes' version, God's mercy seems to have less to do with the illness or the cure than does the character's self-conscious role-playing.
13 A similar case for the character of Hamlet has recently been made. The fragile interiority of Shakespeare's prince at once emerges and recedes from view; its potential for disrupting the social whole is recuperated through the dispersion of its power into other characters who destroy Hamlet's claim to singularity. See Barker 1984.
14 Within the field of Hispanism, the various historical appropriations of Don Quixote have given rise to what in contemporary Cervantes criticism is referred to as the "hard" and the "soft" approach to the novel, the former maintaining that the

original creator of Don Quixote meant to portray him negatively, the latter sympathetically. (The falsity of such a strict dichotomy is a problem for another essay.) Proponents of the hard school, openly concerned with authorial intention, want the novel to be read strictly as burlesque, and accuse the soft or "Romantic critics" of idealizing the hero, relating the text to either the "human spirit" or Spanish history, and/or forcing twentieth-century values on it. See especially Close 1978.

Close's argument itself hinges precariously on notions such as "art is for art's sake," the "individual artistic personality," and "the self-renewing tree of the human spirit," claiming that any reading of the *Quixote* that goes beyond the readily apparent elements of comedy and satire is anachronistic. Terry Eagleton has cited other examples of this critical move in which, for example, we are urged to give up the "tediously zealous search for moral, symbolic and historical meaning in Dickens and acknowledge instead that he is really just rather funny" (1986: 188 n19). Given Close's wideranging disapproval (of any reading which goes beyond whatever Cervantes' initial intentions might have been) and because I am also guilty of the charge Close levels at Américo Castro ("the determination to search 16th [and seventeenth] century Spanish culture with eyes attentive to signs of living relevance to twentieth-century problems"), I am surely a "Romantic critic." At any rate, what I am more interested in trying to understand in the present essay are the ways in which the figure of Don Quixote has been used as a sign for practices which either challenge the hegemonic social groups or seek to protect them.

15 See also Cohen 1985: 146, 149, 234.

16 Lesage's account of the publishing history of Cervantes' *Don Quixote* differs substantially from what modern critics have concluded. Lesage writes in his prologue: "Cervantes, taking it ill that another should build upon his foundation, fell to work again, and published his second part, which he seemed before to have no thoughts of. It is therefore to be observed, that where such things occur in these two second parts, as bear any resemblance to one another, we may easily judge who it was that borrowed from the other, since Cervantes wrote his long after Avellaneda" (Lesage 1705). We are now fairly certain that Cervantes had already written a large portion of his second text before the 1614 publication of Avellaneda's. Whether or not Cervantes may have read the spurious sequel in manuscript form is unknown.

17 Lesage 1705: 435.

18 For a brief reception history of Avellaneda's novel, see Jones 1973. His discussion of the English novelists is especially interesting.

19 Fitzmaurice-Kelly 1905: 18.

20 Tennenhouse 1985: 109–28.

References

Avellaneda, Fernandez de (1971) *El ingenioso hidalgo Don Quijote de la Mancha que contiene su tercera salida y es la quinta parte de sus aventuras*. Ed. Fernando Garcia Salinero. Madrid: Castalia.

——— (1980) *Don Quixote de la Mancha (Part II): Being the spurious continuation of Miguel de Cervantes' Part I*. Trans. A.W. Server and J.E. Keller. Newark, Del.: Juan de la Cuesta.

Barker, Francis (1984) *The Tremulous Private Body: Essays on Subjection*. London: Methuen.

Baroja, Julio Caro (1966) "Honor and Shame: A Historical Account of Several

Conflicts,'' in J.G. Peristiany (ed.) *Honour and Shame: The Values of Mediterranean Society*. Chicago: University of Chicago Press.

Beinart, Haim (ed.) (1974) *Records of the Trials of the Spanish Inquisition in Ciudad Real*. Jerusalem: Israel National Academy of Science and Humanities.

Cervantes (1982) *The Adventures of Don Quixote*. Trans. J.M. Cohen. Harmondsworth: Penguin.

Close, Anthony (1978) *The Romantic Approach of ''Don Quixote'': A Critical History of the Romantic Tradition in ''Quixote'' Criticism*. Cambridge: Cambridge University Press.

Cohen, Walter (1985) *Drama of a Nation: Public Theater in Renaissance England and Spain*. Ithaca, NY, and London: Cornell University Press.

Coto, Sebastian (1639) *Discurso medico y moral de las enfermedades por que seguramente pueden las religiosas dejar la clausura*.

de Vega, Lope (1985) *Fuente Ovejuna*. Ed. Juan Maria Marin. Madrid: Càtedra.

Eagleton, Terry (1986) *Against the Grain: Essays 1975–1985*. London: Verso.

Fitzmaurice-Kelly, James (1905) *Cervantes in England*. Oxford: n.p.

Gilman, Stephen (1951) *Cervantes y Avellaneda: Estudio de una imitacion*. Trans. Margit Frenk Alatorre. Mexico City: Fondo de Cultura Economica.

Jones, Joseph R. (1973) ''Notes on the Diffusion and Influence of Avellaneda's 'Quixote.''' *Hispania* 56 (April): 229–37.

Lesage, René (1705) *A Continuation of the Comical History of the Most ingenious Knight, Don Quixote de la Mancha. By the licentiate Alonzo Fernandez de Avellaneda. Being a third volume; never before printed in English. Translated by Captain John Stephens*.

Mariscal, George (1987) ''Calderon and Shakespeare: The Subject of Henry VIII.'' *Bulletin of the Comediantes* 39: 189–213.

Ponsetti, Helen Percas de (1981) ''Authorial Strings: A Recurrent Metaphor in Don Quijote.'' *Cervantes* 1 (Fall): 51–62.

Riley, E.C. (1962) ''Avellaneda's Quixote,'' in *Cervantes's Theory of the Novel*. Oxford: Clarendon: 212–25.

Tennenhouse, Leonard (1985) ''Strategies of State and Political Plays: A Midsummer Night's Dream, Henry IV, Henry V, Henry VIII,'' in Jonathan Dollimore and Alan Sinfield (eds) *Political Shakespeare: New Essays in Cultural Materialism*. Ithaca, NY, and London: Cornell University Press.

Wurmser, Leon (1981) *The Mask of Shame*. Baltimore and London: Johns Hopkins University Press.

Part Two

Modern Culture: The Triumph of Depth

5

Idleness in South Africa

J. M. Coetzee

I

> The local natives have everything in common with the dumb cattle, barring their human nature. . . . [They] are handicapped in their speech, clucking like turkey-cocks or like the people of Alpine Germany who have developed goitre by drinking the hard snow-water. . . . Their food consists of herbs, cattle, wild animals and fish. The animals are eaten together with their internal organs. Having been shaken out a little, the intestines are not washed, but as soon as the animal has been slaughtered or discovered, these are eaten raw, skin and all. . . . A number of them will sleep together in the veld, making no difference between men and women. . . . They all smell fiercely, as can be noticed at a distance of more than twelve feet against the wind, and they also give the appearance of never having washed.

The above observations on the Hottentots of the Cape of Good Hope come from a compilation assembled from travellers' reports by the Amsterdam publishing house of Jodocus Hondius in 1652, the year of European settlement of the Cape (Hondius 1952: 26–8).[1] By its show of footnoting, its maps, and its engravings of Hottentots in exemplary poses, Hondius's little book means to prove that it is not a work of fantasy: everything it tells of has truly been seen by real people. Within the limits of the veracious, however, the picture it presents is a selective one. The existence of implosives in the language of the Hottentots, their liking for raw entrails, their lax sexual mores, their powerful smell – all these are remarkable facts, selected in the first place by the writers of the original reports from the mass of impressions they received at the Cape for being remarkable, and picked out in turn by Hondius because they seem likely to strike the man in the street in the same way.

In early records of the Hottentots, one finds a repertoire of these remarkable items of fact repeated again and again: their implosives (''turkey-gobbling''), their eating of unwashed intestines, their use of animal fat to smear their bodies, their habit of wrapping dried entrails around their necks, peculiarities of the female pudenda, their inability to conceive God, their incorrigible indolence. Though sometimes these noteworthy items are merely copied from one book to another, we must believe that in some cases they were rediscovered or confirmed at first hand. They constitute some of the most obvious *differences* between the Hottentot and the west European, or at least the west European as he imagined himself to be. But while there are certainly differences, they are perceived and conceived within a framework of samenesses, a framework that derives from the generally accepted thesis enunciated at the opening of the extract from Hondius above: that although the Hottentots may seem like beasts, they are in fact men. Hottentot society being a human society, it can be described within a framework common to all human societies. The categories and subcategories of this framework constitute the samenesses that extend across all societies. They are the universals, while the observations that are filled in under the various headings constitute the *differentia* of societies, which in reports on the Hottentots tend to be the most remarkable *differentia*.

Although the framework of categories within which the travel writers operate is nowhere explicitly set forth by them, it is not hard to extract it from their texts. The list is roughly as follows:

1 Physical appearance
2 Dress: (a) clothing, (b) ornamentation, (c) cosmetics
3 Diet: (a) foodstuffs, (b) cuisine
4 Medicine
5 Crafts: (a) handicrafts, (b) implements
6 Technics
7 Weapons
8 Defence and warfare
9 Recreations
10 Customs
11 Habitation: (a) dwellings, (b) village organization
12 Religion and magic
13 Laws
14 Economy
15 Government
16 Foreign relations
17 Trade
18 Language
19 Character

Though the number of categories may not be nineteen in each case, behind each of the discourses called "Account of the Hottentots" or "Description of the Hottentots" exists some such grid. At its most immediate level, the grid functions as a compositional aid, giving a form of arrangement to the data. But at another level the grid functions as a conceptual scheme, and as such creates the danger that observations may be deformed to fit into one slot or another when they "really" cut across the categories, or that things that belong in no preconceived slot will simply not be seen. Thus – to give hypothetical examples – observations of drug-induced trances and prophecies would fall under Medicine or Religion and magic or (possibly) Law or Government, but not under all four, observations of cattle-slaughtering rituals would fall under Diet or Dress or Religion and magic or (possibly) Economy, but not under all four. Or – to give real examples – O.F. Mentzel ponders over whether the Hottentots' so-called *Pisplechtigheid* is a recreation or a religious ceremony, and adduces it as a proof of the poverty of their language that they use a single term (translated *andersmaken*) to cover the acts of marrying a couple, initiating a youth into manhood, curing an illness, and driving out a spirit (Mentzel 1944, 2: 281, 288). Of course it is too much to expect of the seamen, ship's doctors, and Company officials who contribute to what I will henceforth loosely call the *Discourse of the Cape* that they will put aside their inherited Eurocentric conceptual scheme in favor of a scheme based on native conceptual categories. Aside from the fact that such an expectation would be anachronistic, the collapse of the categories of (say) Diet, Medicine, and Religion and magic into each other threatens a collapse of systematic discourse into what the traveller started with: a series of sightings and observations recollected only because they are striking, remarkable; that is to say, into a mere *narrative* rather than a comprehensive *description*.

The crippling weakness of anthropological narrative as compared with anthropological description is that, yielding its allegiance to chronological sequence, it yields its right to the achronological, spatial, God's-eye organization that belongs to categorical description. Some travel writers try to have the best of both worlds – the immediacy of narrative, the synopticism of description – by disguising the latter as the former. Here, for example, is Christopher Fryke writing of a visit to the Cape in 1685:

> My curiosity led me to enter one of [their huts] and see what kind of life these people led. As I came within, I saw a parcel of them lying together like so many hogs, and fast asleep; but as soon as they were aware of me, they sprang up and came to me, making a noise like turkeys. I was not a little concerned; yet seeing that they did not go about to do me any harm, I pulled out a piece of tobacco and gave it to them. They were mightily pleased, and to show their gratitude they lifted up those flaps of sheepskin which hang before their privy-parts, to give me a sight of them. I made all haste to be gone, because of the nasty stench; also I could readily perceive, that there was nothing special to be seen there. Moreover, some I found at their eating, which made the stink yet more unbearable, since they had only a piece of cowhide, laid out upon the coals a-broiling, and they had squeezed the dung out of the guts, and smeared

> it with their hands over one another. And the hide they take out when it is broiled, and beat it, and so eat it. This so turned my stomach, that I made haste to be gone. (Raven-Hart 1971, 2: 259)

The historical veracity of this narrative is much to be doubted (a few pages later Fryke comes upon a serpent eating a Hottentot). But we may note how the brief story is made up by the stringing together of anthropological commonplaces from the categories Physical appearance, Dress, Diet, Recreations, Customs, Habitation, Language, and Character.

1 The Hottentots sleep by day (idle Hottentot character) in a hut (Hottentot dwelling), lying all over each other (Hottentot sexual mores) like hogs (place of Hottentots on the scale of creation).
2 They make a noise like turkeys (Hottentot language).
3 They accept tobacco (Hottentot recreations) and lift their flaps (Hottentot dress) to exhibit (Hottentot sexual mores) their private parts (reputed anatomical peculiarities of the Hottentots).
4 Fryke is driven away by the stench (Hottentot uncleanliness), observing as he leaves Hottentots smearing one another with dung (Hottentot cosmetics) and eating cowhide and guts (Hottentot diet).

One of the commonplaces of the Discourse of the Cape is that the Hottentots are *idle*. Since idleness is not custom but absence of custom, not recreation but absence of recreation, it usually finds its place in category 19, as part of the "Hottentot character." We find surprisingly little mention of Hottentot idleness in the approximately 150 accounts that Raven-Hart summarizes from travellers who touched at the Cape before 1652.[2] But as the Company began to settle in and accounts of the Hottentots became more detailed, the theme of idleness became more and more prominent, being described and denounced in the same breath:

> "They are lazier than the tortoises which they hunt and eat" – Johan Nieuhof, 1654 (Raven-Hart 1971, 1: 22)

> "They are a lazy and grimy people who will not work. . . . They are idle, and like to sit without doing anything" – Volquart Iversen, 1667 (Raven-Hart 1971, 1: 103)

> "Their chief work is nothing more than to dig up and eat . . . roots. . . . When they are satiated they lie down without a care" – George Meister, 1677 (Raven-Hart 1971, 1: 203)

> "The major work of the men is to lie about, unless hunger drives them" – Johann Schreyer, 1679 (Naber 1931: 40)

> "If they are not hungry, they will not work" – Christopher Fryke, 1681 (Raven-Hart 1971, 2: 234)

"They are very lazy, liking better to go hungry than to work" – Fr.-T. de Choisy, 1685 (Raven-Hart 1971, 2: 269)

"They secure for themselves a luxurious idleness; they never till the soil, they sow nothing, they reap nothing, they take no heed what they shall eat and drink. . . . Whoever wishes to employ them as slaves must keep them hungry" – William ten Rhyne, 1686 (Schapera 1933: 123)

"They are a very lazy sort of people . . . . They choose rather to live . . . poor and miserable, than to be at pains for plenty" – William Dampier, 1691 (Raven-Hart 1971, 2: 385)

"Their native inclination to idleness and a careless life, will scarce admit of either force or rewards for reclaiming them from that innate lethargic humour" – John Ovington, 1693 (Raven-Hart 1971, 2: 396)

"They are extremely lazy, and had rather undergo almost famine, than apply themselves" – François Leguat, 1698 (Raven-Hart 1971, 2: 432)

"They are, without doubt, both in body and mind, the laziest people under the sun. . . . . Their whole earthly happiness seems to lie in indolence and supinity" – Peter Kolb, 1719 (Kolb 1731, 1: 46)

"The men . . . are . . . the laziest creatures that can be imagined, since their custom is to do nothing, or very little . . . . If there is anything to be done, they let their women do it" – François Valentijn, 1726 (Valentijn 1973: 71–3)

"The dull, inactive, and I had almost said, entirely listless disposition, which is the leading characteristic of their minds . . . necessarily produced by the debilitating diet they use, and their extreme inactivity and sloth" – Anders Sparrman, 1783 (Sparrman 1975: 209)

"Lazy, idle, improvident" – O.F. Mentzel, 1787 (Mentzel 1944, 2: 276)

"Perhaps the laziest nation upon earth . . . . [However] the women are very active and industrious in household affairs" – C.F. Damberger, 1801 (Damberger 1801: 57–8)

Though there are occasional dissenting voices,[3] and though the judgments of many writers are based on secondhand evidence or *idées reçues*, one must be struck by the persistence of these strictures, which continue into the period of British occupation of the Cape (see part II below). Idleness, indolence, sloth, laziness, torpor – these terms are meant both to define a Hottentot vice and in the same movement to distance the writer from that vice. Nowhere in the great echo chamber of the Discourse of the Cape is a voice raised to ask whether the life of the Hottentots may not be a version of life before the Fall, in which man is not condemned to eat his bread in the sweat of his brow but instead spends his days dozing in the sun, or in the shade when the sun grows too hot, half-aware of the singing of the birds and the breeze on his skin, bestirring himself to eat when hunger overtakes him, enjoying a pipe of tobacco when it is available, at one with his surroundings and unreflectively contented. The idea that the Hottentot may be Adam is not even entertained for the sake of being dismissed (on the grounds, say, that the Hottentot does not know God). Certainly no one asks whether what looks like Hottentot *dolce far niente* might not be the mere outward aspect of a profound Hottentot contemplative life. At a more practical level, no one asks on what grounds a people whose traditional diet is meat, milk, and *veldkos* should after 1652 decide that gardening is better and begin to till the soil; or why, after artificial appetites for baked bread, tobacco, and spirits have been created in them, they should want to sell any more of their labor than is required for the immediate satisfaction of these appetites. No one bothers to put, save rhetorically, the ethical question of which is better: to live like the ant, busily storing up food for winter, or like the grasshopper, singing in the sun all day, heedless of the morrow. No one even reflects that the Hottentot, with his meagre possessions and his easily satisfied wants, shows a way to escape the cares of civilization.

It is not enough to answer the question of why questions like these were not asked by saying that the kind of person responsible for the Discourse of the Cape would never have thought of asking them. Certainly many of the travel writers were straightforward Company officials, ship's captains or military men; but there were also scientists of distinction among them (Kolb, Sparrman) as well as men of learning (Ten Rhyne) and serious observers (Schreyer). Furthermore, in Europe the fabled Hottentot did succeed in becoming a term in philosophical debate about Man, though that was less a debate about the natural state of man than about whether there was a single creature called Man or several races of men, some nearer to the beasts than others.[4] To understand why the Hottentot way of life, characterized by (and stigmatized for) its idleness, was in no way held up to Europe as a model of life in Eden, we must in the first place gain some perspective on ideas about idleness reigning in Europe at the time when Europe, and particularly Protestant Europe, was learning about the Hottentot.

We may begin by recalling that the contemplative life, validated by the medieval Church as a higher kind of life, and indeed a higher kind of activity, than the active life, was rejected by Luther as part of his rejection of a privileged spiritual status for the clergy. In Germany of post-Reformation times, in particular, preachers placed increasing emphasis on *work* as the fundamental divine edict that all men must obey to atone for Adam's fall. To be idle was to disobey this edict, and to be improvident – to depend on God's providence to save one from starving – constituted a further offense, a tempting of God. The devotional books of the period thunder against the ''curse of idleness''; the community of Herrnhut, founded in 1727 and destined to form the model for the mission of the Moravian

Brethren to the Hottentots of Africa, is representative of the age in writing into its statutes the requirement that everyone who joins the community must labor for his own bread (Vontobel 1946: 67–70, 38: Marais 1957: 147–8).

As part of the Reformation, too, the Renaissance ideal of personal leisure, *otium*, time for self-cultivation – an ideal resurrected from classical times – was rejected. The distinction between base idleness and leisure as a higher form of idleness fell away: mankind was held to be so weak that without the discipline of continual work it was bound to relapse into sin. Bucer suggested excommunication as the ultimate remedy for idleness (Vontobel 1946: 78). In Calvinism, in particular,

> waste of time [became] . . . the first and in principle the deadliest of sins. The span of human life is infinitely short and precious to make sure of one's own election. Loss of time through sociability, idle talk, luxury, even more sleep than is necessary for health . . . is worthy of absolute moral condemnation. (Weber 1976: 157–8)

At the same time a war on social parasitism was set in train; even almsgiving was condemned as "a great sin" in that it encouraged people to evade God's edict on work (Vontobel 1946: 75). By the middle of the seventeenth century what Michel Foucault calls "the great confinement" had got under way as the culmination of a series of measures to put an end to vagrancy and begging as a way of life, beginning with the confinement of the beggar class and going on later to sweep up the insane and the criminal. During crises of unemployment the houses of confinement became in effect prisons for the workless; during economic upswings they acted as hostels cum factories. As productive organizations they were a failure, but that did not matter. Their prime purpose was not to show a profit but to proclaim the *ethical* value of work. In this earliest phase of industrialization and this primitive phase of economic thinking, as Foucault points out, labor and poverty were held to be simple polar opposites: labor was imagined as having the power to abolish or overcome poverty "not so much by its productive capacity as by a certain force of moral enchantment" (Foucault 1967: 48–55).

Though conducted with greater ferocity in Protestant countries, the war on beggars took place in both Catholic and Protestant Europe and continued as long as vagrancy remained a significant social problem, well into the nineteenth century. The anathema on idleness, which was part of this war, did not falter with the Enlightenment; for the revolt against authority that constituted the Enlightenment simply replaced the old condemnation of idleness as disobedience to God with an emphasis on work as a duty owed by man to himself and his neighbor. Through work man embarks on a voyage of exploration whose ultimate goal is the discovery of man; through work man becomes master of the world; through a community of work society comes into being. Karl Marx is wholly a child of the Enlightenment when he writes:

"The entire so-called history of the world is nothing but the creation of man through human labour" (Marx 1975: 305).

Both of these attitudes towards idleness – that idleness is sin, that idleness is self-betrayal – can be seen in the Discourse of the Cape. In the first hundred years or so of settlement, the idleness of the Hottentots is denounced in much the same spirit as the idleness of beggars and wastrels is denounced in Europe. One might even say that the rhetoric used to justify a class war in Europe is unthinkingly transferred to the colonial situation to condemn the refusal of the natives to be drawn into the economy as wage laborers. This formulation must be qualified, however, for the first wave of denunciation of Hottentot idleness belongs not so much to the discourse of the *rulers* of the Cape, where one might expect it if the problem of Hottentot labor were uppermost, as to the rudimentary ethnographic discourse of travel literature.[5] Furthermore, if one is to be logical, idleness is precisely what the newly-arrived colonizer might expect in a heathen folk who had not heard of God's word and knew nothing of the ban on idleness. In fact, the disguising of an attack on what we may call Hottentot passive resistance to wage labor as a denunciation of idleness belongs to a later stage of the history of the Cape; the highlighting of Hottentot idleness, though understandable in men with seventeenth-century Protestant backgrounds is a response to more immediate frustrations. We can move closer to an understanding of what the idleness of the Hottentots means to the early ethnographers if we ask what it is that the Hottentots are *not doing* when it is asserted that they are idle.

We can begin to approach an answer if we note that the charge of idleness usually comes together with, and sometimes at the climax to, a set of other characterizations: that the Hottentots are ugly, that they never wash but on the contrary smear themselves with animal fat, that their food is unclean, that their meat is barely cooked, that they wear skins, that they live in mean huts, that male and female mix indiscriminately, that their speech is not like that of human beings. What is common to these charges is that they mark the Hottentot as *underdeveloped* – underdeveloped not only by the standard of Man. If he were to develop dietary taboos, ablutionary habits, sexual mores, "processed" rather than "wild" clothing, a varied body decoration rather than a uniform coating, domestic architecture and technology, a language rather than animal noises, he would become, if not a Hollander, at least more fully Man. And the fact that he self-evidently does not employ his faculties in developing himself in these ways, but instead lies about in the sun, is proof that it is sloth that must be blamed for holding him back from this becoming.

Yet what kind of creature is the Man that the Hottentot, who in his present state "could be accounted more among the dumb beasts than among the company of reasoning men" (J.C. Hoffman 1680, in Naber 1931: 31), refuses out of idleness to become? It is Man with a developed Physical appearance, Dress, Diet, Medicine, Crafts, etc. – in other words, what we may call Anthropological Man. The Hottentot is Man but not yet Anthropological Man; and what keeps him in his backward state is idleness. Thus his idleness has the status of an *anthropological scandal*: despite the fact that nothing remoter and more *different* from European Man can be imagined than the Hottentot, the Hottentot generates an extremely impoverished set of differences to inscribe in the table of categories. Where he ought to be generating data for the categories, he is merely lying about. Where he ought to have Religion, there is a virtual blank. His Customs are casual. His Government is rudimentary. Though far more *different* from the European than the Turk or the Chinese is, the Hottentot paradoxically presents far fewer *differences* for the record.

The force of the righteous condemnation that the Discourse of the Cape brings to bear on the Hottentot comes from the accumulated weight of two centuries of denunciation of idleness, from the pulpit and the judicial bench, in Europe. But his idleness is felt with particular animosity by the travel writer, the proto-anthropologist, to whom he promises so much in the way of difference and yields so little. It is striking that, once we move out of the categorical discourse of anthropology, where the scheme requires the writer to inscribe eighteen or nineteen blocks with lists of remarkable differences, to the discourse of history, where at its simplest the form requires the writer merely to chronicle each day the remarkable happenings of that day, there is far less stress on the idleness of the Hottentots (see note 5). Indeed, in history the Hottentots suddenly seem all too busy, intriguing with one another, driving off cattle, begging, and spying.

I am far from wanting to deny that, in so far as the word *idle* has any objective meaning, the Hottentots were idle, or to assert that the condemnation of Hottentot idleness had nothing to do with the desire of the colonists to impress them as laborers. What I do wish to stress, however, is that the most universal denunciation we find among travel writers represents a reaction to a challenge, a scandal, that strikes particularly near to home to them *as writers*: that the laziness of the Hottentot aborts the most promising of discourses about elemental man. Nor are these early writers the last to respond with frustration to the recalcitrance of the colonies to generate materials to fill out their inherited forms. The ethnographer Gustav Fritsch, travelling around South Africa in the 1860s, observes that it would not be possible to use Boor life as material for stories because nothing ever happens (Fritsch 1868: 161); and, at much the same time, Nathaniel Hawthorne is lamenting that the ''commonplace prosperity'' without surprises and reversals, the ''broad and simple daylight'' that he finds in America make an American novel impossible (Hawthorne 1961: vi). In each case the colonial material is condemned as inadequate, too exiguous for the European form; and in each case we might ask whether the new materials do not require a recasting of the old forms, the old conceptual frameworks. The moment when the travel writer condemns the Hottentot for doing nothing marks the moment when the Hottentot brings him face to face (if he will only recognize it) with the limits of his own conceptual framework.

That we are faced here with a phenomenon of more significance than the failure of a set of casual writers with rough, workaday minds to escape their ethnocentric prejudices, can be seen if we turn to a major anthropological work, Jean-Jacques Rousseau's *Discourse on the Origin of Inequality* (1754). In a paragraph in which he specifically mentions the Hottentots, Rousseau characterizes Man in his savage state as "solitary, indolent [*oisif*], and perpetually accompanied by danger," a creature who "cannot but be fond of sleep." He is lifted out of this state by the invention of tools, which bring about the first revolution in human culture and permit him an easier, less dangerous life. In this new phase of comparative leisure (*loisir*) he begins to create conveniences for himself that eventually develop into the yoke of civilization. This phase, intermediate between savage indolence and the second cultural revolution that will come with the rise of metallurgy and agriculture, will introduce private property and social inequality, and will make work an unavoidable part of daily life, is singled out by Rousseau as "the happiest and most stable of epochs."

> The example of savages, most of whom have been found in this state, seems to prove that men were meant to remain in it, that it is the real youth of the world, and that all subsequent advances have . . . in reality [been steps] towards the decrepitude of the species. (Rousseau 1913: 169, 195–9)

Yet when Rousseau comes to spell out what life is like in this "happiest and most stable of epochs," the description he gives, though based on reports of the New World, could very well be a panorama of Hottentot life: rustic huts, clothes made of skins, adornments of feathers and shells, bows and arrows as weapons, clumsy musical instruments. What, then, is the crucial difference that prevents the Hottentot from belonging to the golden age? Certainly his unsavory personal habits and his infringement of European taboos on the preparation and consumption of meat play their part. But the essential difference is that the Hottentot is *indolent*, he spends his "free" time sleeping, while among savages after the tool-making revolution "free" time becomes leisure, time devoted "industriously" to the elaboration of "conveniences" (ibid.). Rousseau thus, in line with Enlightenment thought, resurrects the humanistic opposition of *leisure* (Roman *otium*, Greek *schole*, time for self-improvement) to *idleness*: the Hottentot does not belong to the happiest of epochs because he is idle.[6] Leisure holds the promise of the generation of all those differences that constitute culture and make man Anthropological Man; idleness holds no promise save that of stasis.

II

Condemning the Hottentot for his idleness, the early Discourse of the Cape effectively excludes him from Eden by deciding that, though he is human, he is not in the line of descent that leads from Adam via a life of toil to civilized man. The Hottentot, that is to say, is not an *original* of civilized man. Although one cannot postulate that so farsighted an intention lay behind the strictures of the early writers, this conclusion nevertheless prepares the ground for the next phase of the attack on the Hottentot way of life, a way of life that, even as early as the mid-eighteenth century, was not confined to ethnic Hottentots but had its adherents among the remoter Dutch Boors. Thus O.F. Mentzel, who spent the years 1732–41 at the Cape, writes that some of the Boors

> have accustomed themselves to such an extent to the carefree life, the indifference, the lazy days and the association with slaves and Hottentots, that not much difference may be discerned between the former and the latter. (Mentzel 1944, 2: 115)

This "Hottentot" life of idleness and improvidence, this *lekker lewe*, never wins a spokesman in the Discourse of the Cape. Its only chance of making its voice heard would have been to assert an analogy between the Hottentot and unfallen man, between the Cape and Eden, and via this analogy to claim its legitimacy. But, for reasons that we have explored, its voice is silent, or nearly so,[7] and though the idle life continues to be lived on all sides, it does so illegitimately, defensively, and invariably, when disclosed in print, as a scandal.

The indolence of the Hottentots is discovered afresh by British commentators after 1795. Robert Percival writes of "the peculiar indolence and want of vigour of the Hottentot character," which he diagnoses as "an original bad quality" (Percival 1804: 84–5). Barrow writes of indolence as "the principal cause of [the] ruin" of the Hottentots, "a real disease, whose only remedy seems to be that of terror," the remedy of hunger being inadequate (Barrow 1806, 1: 102). Burchell (1822) praises the Moravian missionaries for their insistence on manual labor and predicts that once they have taught the Hottentots "the necessity of honest industry" they will have "cut off the root of, at least, half the miseries of the Hottentot race" (Burchell 1953, 1: 80). The consensus is that the Hottentot way of life, characterized by low-level subsistence maintained by a minimal resort to wage labor ("laziness"), wandering in search of greener pastures ("vagrancy"), and a sometimes casual attitude toward private property ("thieving"), will have to be reformed by *discipline* (a key word of the age) if the Hottentot is to have any stake ("pull his weight") in the Colony.

To the extent that it recognizes the fact that a Hottentot tribal life within the areas settled by colonists no longer exists, and that the only viable future for the Hottentots is within the colonial economy, this attitude can be regarded as hard-headed. But to the extent that it sees ''indolence'' as part of the Hottentot racial ''character,'' an ''original bad quality'' that only generations of strict discipline will eradicate (''cut off the root of''), we can accurately call it a racist attitude. It is entirely unreflective, seeing only squalor, disease, and blank idleness, shutting its eyes where it might also have seen a way of existing for which people might settle for the sake of avoiding the greater wretchedness of lifelong manual labor. In contrasting inherent European diligence with inherent Hottentot sloth, it seems to have forgotten the history of the early phase of industrialization in Europe, where it required a reformation of ''character'' occupying generations before the laboring class would embrace the principle that one must work harder than is required to maintain the level of material existence one is born into.[8] To bring about this reformation, to make people believe that ''the opportunity of earning more was [more] attractive than that of working less,'' required a sustained program of ideological indoctrination conducted through schools, churches, and the popular press, a program meant to convince the lower classes that work was ''necessary and noble'' (Anthony 1977: 41, 22). A writer like Barrow, son of a self-made man and influential advisor on colonial policy, is wholly committed to this ideology, as were the missionaries to whom the conduct of the program of indoctrination was consigned in the colony. To persuade the Hottentots of Bethelsdorp to spend their time collecting the juice of aloes, John Philip of the London Missionary Society allowed a shop to be opened on the mission station. The ''experiment'' of getting the Hottentots to work by holding before them the temptation of desirable articles for purchase succeeded: ''money instantly rose in estimation among them.'' Whether what Philip calls ''the creation of artificial wants'' was an appropriate one for missionaries to undertake is not to the point here. To Philip as a social thinker, the Hottentot had no future unless he learned to sell his labor. As Philip candidly put it to his charges, they could not hope to use the mission stations as refuges from the dragnet of colonial authorities trying to settle them as serfs on farms, treating them as pockets of land where a pre-colonial regime of idleness, improvidence, and easy morals might be maintained: quite aside from the fact that the missionaries would not sanction such a way of life, ''the world, and the Church of Christ'' (which funded the missions) ''looked for civilization and industry as proofs of [the Hottentots'] capacity for improvement . . . [for] men of the world had no other criteria by which they could judge'' (Philip, 1828, 1: 204–5, ix, 212). In other words, if the Hottentots did not learn to work on the mission stations, the mission stations would close and they would be left to the mercies of the farmers. One way or the other, work they must. Thus, while the colonists were denouncing the LMS stations as ''nests of idleness,'' idleness was precisely what the missionaries saw as the prime feature that had to eradicated from the Hottentot ''character.''[9] If the LMS stations never became quite the hives of industry that the Moravian stations were reputed to be, it was largely because they did not practice the exclusion of people who came not to work but to share in the prosperity of their kinsfolk. As one observer lamented, the more industrious Hottentots of the Kat River Settlement were having their wealth eaten up by "squatters" (presumably relatives) who "indulged in habitual sloth and listless inactivity"; and John Philip had similar observations to make about the mission stations, where "the means of the industrious [are] eaten up by the idle" (Marais 1957: 225, 249).

But the true scandal of the nineteenth century was not the idleness of the Hottentots (by now seen as inherent in the race) but the idleness of the Boors. The sliding of farmers into an idle way of life can be traced back to the first decades of settlement: governor Wagenaar, Van Riebeeck's successor, wrote to the Chamber in 1663 suggesting that half a dozen of the free farmers ought to be called home because of their "indolence and . . . irregular and debauched lives"; from the Chamber, familiar with the problem from the Indies, he received the tolerant reminder that "our people, when abroad, are at all times with difficulty induced to work," and the suggestion that he should rely more on slaves (Moodie 1960: 270, 279). "Too much good fortune hath bred sloth among the farmers," says Gravenbroek in 1695 (Schapera 1933: 273). A century later Le Vaillant comments on the frontier farmers that "from the profound inaction in which they live, one would suppose their supreme felicity to consist in doing nothing."[10]

Not only the farmer but the burgher of Cape Town was afflicted with this lapse into sloth. Stavorinus describes a day in the life of a burgher at the end of the eighteenth century: a long smoke and stroll in the morning, an hour or two of business, a midday meal followed by a snooze, an evening of cards – all in all "a very comfortable life" (Stavorinus 1796: 248). Percival and Barrow confirm his account a decade later: "a most lamentable picture of laziness and indolent stupidity," Percival calls it (Percival 1804: 255; Barrow 1806, 2: 100–1).

The harshest remarks of nineteenth-century commentators are, however, reserved for the Boors of the frontier. In his survey of the productive potential of the Colony, Barrow says:

> Luckily, perhaps, for them, the paucity of ideas prevents time from hanging heavy on their hands. . . . [A] cold phlegmatic temper and [an] inactive way of life . . . indolence of body and a low groveling mind

Seeing sloth as by now part of the "nature" of the Boor, he suggests that the Colony will not become productive until this "nature" is changed, or, failing that, until the Boors are replaced by more industrious and enterprising settlers (Barrow 1806, 1: 32; 2: 118; 1: 386).

The refrain is taken up by every traveller who penetrates into the back country and encounters farmers living on huge tracts of land in mean dwellings, barely literate, rudely clad, surrounded by slaves and servants with too little employment, disdainful of manual labor, content to carry on subsistence farming in a land of potential plenty. Percival comments:

> There is I believe in no part of the world an instance to be found of European adventurers so entirely destitute of enterprise, and so completely indifferent to the art of bettering their situation.

The women of the frontier he finds particularly "lazy, listless and inactive," a judgment confirmed by Moodie: they are "exceeding torpid and phlegmatic in their manners and habits, dirty and slovenly in their dress" (Percival 1804: 211, 205; Moodie 1835, 1: 170). On the frontier "days and years pass on in miserable idleness," says Campbell (Campbell 1815: 81). Burchell observes that the new immigrant, full of enterprise and energy, swiftly rises to prosperity; but then "he adopts the rude manners [of the Africander] he at first despised, and, step after step, his life degenerates into mere sensual existence." Burchell repeats Barrow's diagnosis that sloth has become part of the Boor character, and follows Barrow's prescription that some kind of missionary work is necessary to bring the Boor into the modern world:

> The ease of an indolent life, with all its losses, is so much more agreeable to [them] than the labour of an industrious one with all its advantages, that the minds of such men must be entirely new-modelled before they can be capable of receiving the improvements of other countries. (Burchell 1953, 1: 194, 377)

Fifty years after Burchell, Gustav Fritsch finds among the Boors "a degree of indolence and indifference" that is "absolutely Chinese in its constancy" (Fritsch 1868: 89–90); and the Carnegie Commission of the 1930s notes again the "indolence" of the "poor white" descendants of these farmers, an indolence that it ascribes to, *inter alia*, the South African climate, prejudice against "kaffir work," and a tradition of easy existence (Wilcocks 1932: 52–79).

The squalor, illiteracy, and above all sloth of Boor life dismays the apostles of colonialism because it constitutes sinister evidence of how European stock can regress after a few generations in Africa.[11] The Boor further betrays the colonizing mission by settling down to scratch nothing but a bare living from the soil, since to justify its conquests colonialism has to show that the colonist is a better steward of the earth than the native (the text usually cited in support of this argument is Matthew 25: 14–30, the parable of the talents). Nor can one neglect the element of chauvinism in the comparisons British commentators draw between the diligent English yeoman and the listless Dutch Boor.

But there is a further component to the British response to Boor idleness a component of moral outrage stemming from the realization that Boor ease is achieved at the expense of the misery of slaves and servants. The ease of the farmer is scandalous because it is corrupt: the case of the Cape Colony confirms the dictum going back to antiquity (see Davis 1966: ch. 3; Lecky 1869, 1: 277), that slaveholding corrupts the slaveholder. "The possession of slaves, and the subjection of the Hottentots . . . have been the source of the greatest demoralization of all classes in this colony" (Moodie 1835, 1: 176). "The taint of slavery, here as elsewhere makes the white man lazy" (Alexander 1838, 1: 70). In the Cape the taint works in a particularly insidious way because, while on the one hand the fact that manual work is the province of slaves creates a prejudice against manual work for the master, idleness as a pervasive way of life has the consequence that about each farmer-patron clusters a band of dependants and hangers-on doing little work and getting the poorest wage in return. Thus while a disdain for work becomes institutionalized among the masters, the system does not even have the compensation that habits of industry are fostered among the servants, who often prefer Boor masters to British, since the latter, while they pay more, demand too much work (Marais 1957: 130–1).

On the other hand, the idleness of the Boors does not create the same crisis for the commentator *as writer* that the idleness of the Hottentot created in the seventeenth century. For while the framework of the earlier writing was that of a nascent science of Man, with its universal cultural categories, the typical nineteenth-century commentary is a loose narrative in which the narrator is free to move across the face of the Colony, seeing sights, having hunting adventures, meeting new people, recording anecdotes and oddities. In fact, the genre is intentionally miscellaneous, a *causerie*, as the extended chapter summaries indicate.[12] Almost any material is fit to fill the ethnographic space left by Hottentot and Boor inactivity, as long as it is entertaining.

The fact that Boor idleness is achieved at the expense of a servile class, and is therefore in a crucial respect different from the old Hottentot idleness, has the natural consequence that the philosophical question that did not get asked regarding the Hottentots gets asked all the less regarding the Boors, namely: if we look beneath the dirty skins, the clouds of flies, and the rude clothing, do these frontier farmers not stand for a rejection of the curse of discipline and labor in favour of an African way of life in which the fruits of the earth are enjoyed as they drop into the hand, work is avoided as an evil, and leisure and idleness become the same thing? The moral and political outlook of nineteenth-century British commentators makes it unlikely that they will entertain such an idea favourably. Nevertheless, the fantasy of an African Eden does not get suppressed entirely, particularly once the efforts of the first wave of literary Romanticism to locate unfallen man in the child or the peasant or the savage have made the quest for man's origins a commonplace theme in travel writing. No one asks whether the torpor of the Hottentot or the sloth of the Boor is a sign that all wants have been met, all desires have been stilled, and Eden has been recovered. But there is a revealing moment in the *Travels* of Burchell, perhaps of early-nineteenth-century commentators the most readily sympathetic to native ways of life. In 1812 Burchell spent an evening among a group of bushmen somewhere between Prieska and De Aar annotating their music and watching their dancing. At midnight he retired to bed. Of the evening he says:

> Had I never seen and known more of these savages than the occurrences of this day, and the pastimes of this evening, I should not have hesitated to declare them the happiest of mortals. Free from care, and pleased with little, their life seemed flowing on, like a smooth stream gliding through flowery meads. Thoughtless and unreflecting, they laughed and smiled the hours away, heedless of futurity, and forgetful of the past.

Though hedged around with conditions (''Had I never seen and known more . . .''), this is a vision of man before the Fall, a vision whose seductions Burchell acknowledges:

> I sat as if the hut had been my home, and felt in the midst of this horde as if I had been one of them; for some moments . . . forgetting that I was a lonely stranger in a land of wild untutored men. (Burchell 1953, 2: 48)

Even Barrow, so censorious of the Boors, so dismissive of the Hottentots (''perhaps the most wretched of the human race,'' Barrow 1806, 1: 93), can be impressed by the kaffirs and speculate that their natural nobility of bearing must be the consequence of simple diet, regular habits, abstinence from alcohol, pure air, plenty of exercise, and chastity – in other words, the consequence of a freedom from the more debilitating features of civilization which the British public school would later try to reproduce. But it is worth noting that what draws Barrow to these African Spartans is in the first place the ''sprightliness, activity and vivacity'' of their appearance (Barrow 1806, 1: 119), while Burchell's eulogy of the bushmen comes only after they have danced ''till morning light announced that other duties claimed their time'' (Burchell, ibid.). The savage thus has to bend his neck to the yoke of ''activity,'' ''duties,'' before the thought may be admitted that he belongs to a Golden Age.

III

Today we are not likely to be so censorious of the idleness of the Hottentots (the Boors perhaps present a somewhat different case). We have a century of anthropological and historical discipline behind us to make us wary of observing the lives of foreign peoples too cursorily and from too self-centred a viewpoint. Given a chance to visit the seventeenth-century Cape, we might reasonably expect to see features of Hottentot life that seventeenth-century observers missed. We might be more sensitive to seasonal variations in activity and to the rhythm of a Hottentot ''week.'' We might be more hesitant about calling an entire people lazy on the grounds that the men lie around while the women do the work (a fact that several early travellers noted). Regarding diet, we might concentrate less on meat and fish and more on the gathering activities of women, with a greater consciousness of the role that strange dietary customs play in the formation of hostile prejudice.[13] We might be more cautious of taking those Hottentots on the fringes of Dutch settlement as typical of all Hottentots. With our wider historical perspective, we might also appreciate better what a massive cultural revolution must occur when a people moves from a subsistence economy to an economy of providence and from pastoralism to agriculture, a move, indeed, in which the notion of *work* may be said to make its appearance in history.

Yet in the very open-mindedness we might like to imagine extending toward the Hottentot from the modern science of Man lies the germ of an insidious betrayal of the Hottentot. For, no less than in the science of Man that met and was frustrated by the real Hottentots, the modern science of Man has at its foundation a will to see a culture *at work* in a society. The science of Man is itself a discipline, one of what Foucault calls the disciplines of surveillance; among its tasks are the tracking down and investigating of obscure societies in all quarters of the globe, the photographing and recording and deciphering of their activities (Foucault 1977: 224). If the Hottentot did not absorb the ideology of work in a generation, we cannot expect the western bourgeois to drop his allegiance to it in a day. It would be particularly rash to expect the modern researcher and writer to respond more generously than his ancestors to a way of life so indolent that, in its extreme form, it presented him with *nothing to say*. The temptation to claim that there is something *at work*, when there is nothing, is always strong. The present essay does not entirely resist that temptation. The challenge of idleness to work, its power to scandalize, is as radical today as it ever was. Indeed – though it takes us outside the bounds of the present discussion – we might wonder whether the challenge that idleness presents to all philosophical enterprise is any less powerful or subversive than the challenge presented by eroticism, in particular by the *silence* of eroticism (I refer here to the work of Georges Bataille – see Bataille 1977: 273–6).

The history of idleness in South Africa cannot be dismissed as a side issue or a curiosity. One need only look around at the face of South African labor to confirm this. The idleness of the Boor is still with us in "job reservation," in taboos on certain grades of manual work ("hotnotswerk," "kafferwerk"), and in rituals of leisure indistinguishable from idleness (sitting on the stoep, lying on the beach). The idleness of the native is still with us in a tradition of overemployment and underpayment, maintained from both sides of the fence, in terms of which two men are hired to do one man's work, each working half the time and standing idle half the time, each getting half of one man's wage. The luxurious idleness of the settler is still denounced from Europe, the idleness of the native still deplored by his master. I hope it is clear that the present essay by no means adds its voice to the chorus of moralizing disapproval. On the contrary, I would hope that it opens a way for the reading of idleness since 1652 as an authentically native response to a foreign way of life, a response which has rarely been defended in writing, and then mostly evasively (one thinks of H.C. Bosman), but which has exerted a powerful attraction since the days when commentators began to shake their heads over Europeans who, from too much intercourse with Hottentots, were denying their birthright and sinking into a life of sloth. It is a measure of how powerful this attraction has remained into the twentieth century that after 1948 the authorities embarked – and, to the extent that they were responding to social realities, found themselves compelled to embark – on a program of laws to reform South African society. Two cornerstone measures of this program were the Immorality Act and the Mixed Marriages Act, laws whose primary intention and whose practical effect it was to take away the freedom of white men to drop out of the ranks of the labouring class, take up with brown women, settle down to more or less idle, shiftless, improvident lives, and engender troops of ragged children ranging in hue from dark to light, a process that, if allowed to accelerate, would in the end spell the demise of White Christian civilization at the tip of Africa.

Notes

1 In this essay, the words *Hottentot*, *Boor*, *Kaffir*, and *bushman* are used throughout to accord with the usage of the writers under discussion.

2 Only three travelers mention idleness, and all three *deduce* the idleness of the Hottentots from the fact that they do not practice agriculture rather than *observing* it: Edward Terry, 1616; Augustin de Beaulieu, 1622; and Johan Wurffbain, 1646 (Raven-Hart 1967: 83, 101, 165).

3 See, for example, Grevenbroek, 1695, in Schaperra 1933: 271–3.

4 François Bernier (1620–88) concludes that the Hottentots are "a different species" from the Negroes of Africa. William Petty calls them "the most beastlike of all the souls of men." John Locke (1632–1704), however, suggests that the intellect of the Hottentot only seems "brutish" because of environmental influences on him. Buffon (1707–88) argues that the distance between the Hottentot and the ape is far greater than the distance between the Hottentot and the rest of mankind. Johann Blumenbach (1752–1840) argues that, while the Hottentot may seem to belong to "a different species," there is in fact only "one variety of mankind." See Slotkin 1965: 95, 90, 173, 184, 189.

5 Van Riebeeck alludes to the idleness of the Hottentots only once, in a dispatch to the Chamber dated 14 April 1653, where he begs to be removed from among these "dull, stupid, lazy, stinking people" to Japan, where his talents might be of some use. His *Journal* contains no reference to Hottentot idleness. Though his successors, Wagenaar and Borghorst, have more than a little to say about idleness, it is the idleness of the free farmers they condemn. See Van Riebeeck 1952; Moodie 1960: 32, 270, 294, 304.

6 On the classical background to the notion of leisure, see De Grazia 1962: 11–25.

7 For one exception to all the head-shaking, see Simon de la Loubière, 1687: "In such poverty [the Hottentots] are always gay, always dancing and singing, living without occupation or toil" (Raven-Hart 1971, 2: 269).

8 Max Weber: "A man does not 'by nature' wish to earn more and more money, but simply to live as he is accustomed to live and earn as much as is necessary for that purpose. Wherever modern capitalism has begun its work of increasing productivity of human labour by increasing intensity, it has encountered the immensely stubborn resistance of this leading trait of pre-capitalistic labour" (Weber 1976: 60). See also Hutt 1939: ch. 5, "Preferred Idleness."

9 "Nests of idleness" is the phrase used by the colonial magistrates in 1849 in their denunciation of the mission stations (Marais 1957: 197). For the comments of Revd John Campbell, inspector of the LMS stations, on the "idleness and sloth" of Hottentots who come to the stations from both kraals and farms, see Campbell 1815: 92–3. See also Burchell's report that the missionaries at Klaarwater continually complained of "the laziness of the Hottentots" (Burchell 1953, 1: 246).

10 Le Vaillant 1796, 1: 59. See also Paterson 1789: 84.

11 Degeneracy was already held in prospect by Mentzel in 1787. Writing of those Boors who "prefer to live in the most distant wilderness among the Hottentots," he expresses his fear that if they do not intermarry with new European stock they will "degenerate and become uncivilized," like the Scots or Wends or Scythians: already "their nature is wild, their education bad, their thoughts base and their conduct ill-bred" (Mentzel 1944, 2: 120).

12 A typical chapter summary: "Wesleyville – Its delightful scenery – Second and third missionary stations – Interpreters and guides – Anecdotes of the elephants – Strange scene – Hottentot eloquence – Grave argument – Artifice – Criticism, and humour – Games – Evening amusements – Shooting hippopotami – The River Kei – The Incagalo – Kaffir chief and his staff – Anecdotes" (Rose 1829: x).

13 Marshall Sahlins describes "the characteristic paleolithic rhythm of a day or two on, a day or two off" (Sahlins 1974: 23). Richard B. Lee, writing of the Dobe bushmen, observes how surprisingly high a proportion of their food intake comes from vegetable foods collected by women (Lee 1968: 33). Richard Elphick discusses the effect of "radically foreign" customs on the formation of European prejudice against the Hottentot (Elphick 1977: 193–200).

References

Alexander, James Edward (1838) *An Expedition of Discovery into the Interior of Africa*. 2 vols. London: n.p.

Anthony, P.D. (1977) *The Ideology of Work*. London: Tavistock.

Barrow John (1806) *Travels into the Interior of Southern Africa*, 2nd edn. 2 vols. London: n.p.

Bataille, Georges (1977) *Death and Sensuality*. Trans. anon. New York: Arno.

Burchell, William J. (1953) *Travels in the Interior of Southern Africa*. 2 vols. London: Batchworth.

Campbell, John (1815) *Travels in South Africa*, 3rd edn. London: n.p.

Damberger, C.F. (1801) *Travels in the Interior of Africa . . . 1781–97* Trans. anon. London: n.p.

Davis, David Brion (1966) *The Problem of Slavery in Western Culture*. Ithaca, NY: Cornell University Press.

De Grazia, Sebastian (1962) *Of Time, Work and Leisure*. New York: Twentieth Century Fund.

Elphick, Richard (1977) *Kraal and Castle: Khoikhoi and the Founding of White South Africa*. New Haven: Yale University Press.

Foucault, Michel (1967) *Madness and Civilization*. Trans. Richard Howard. New York: New American Library.
——— (1977) *Discipline and Punish*. Trans. Alan Sheridan. New York: Pantheon.
Fritsch, Gustav (1868) *Drei Jahre in Süd-Afrika*. Breslau.
Hawthorne, Nathaniel (1961) *The Marble Faun*. New York: New American Library.
Hondius, Jodocus (1952) *A Clear Description of the Cape of Good Hope*. Trans. L.C. van Oordt. Cape Town: Van Riebeeck Festival Book Exhibition Committee.
Hutt, W.H. (1939) *The Theory of Idle Resources*. London: Cape.
Kolb, Peter (1731) *The Present State of the Cape of Good Hope*. Trans. Medley. 2 vols. London.
Lecky, William (1869) *History of European Morals*. 2 vols. London.
Lee, Richard B. (1968) "What Hunters Do for a Living," in Richard B. Lee and Irven DeVore (eds) *Man the Hunter*. Chicago: Aldine: 30–48.
Le Vaillant, François (1796) *New Travels in the Interior Parts of Africa*. Trans. anon. 3 vols. London.
Marais, J.S. (1957) *The Cape Coloured People, 1652–1937*. Johannesburg: Witwatersrand University Press.
Marx, Karl (1975) "Economic and Philosophic Manuscripts of 1844," in Karl Marx and Friedrich Engels, *Works*, 3. Trans. Clemens Dutt. London: Lawrence & Wishart: 229–346.
Mentzel, O.F. (1944) *A Geographical and Topographical Description of the Cape of Good Hope*. Trans. G.V. Marais and J. Hoge. Ed. H.J. Mandelbrote. 2 vols. Cape Town: Van Riebeeck Society.
Moodie, J.W.D. (1835) *Ten Years in South Africa*. 2 vols. London: n.p.
Moodie, Donald (ed. and trans.) (1960) *The Record: or a Series of Official Papers relative to the Condition and Treatment of the Native Tribes of South Africa*. Cape Town: Balkema.
Naber, S.P. L'Honoré (ed.) (1931) *Reisebeschreibungen von deutschen Beamten und Kriegsleuten im Dienst der Niederländischen West- und Ost-Indischen Kompagnien, 1702–1797*, 7. The Hague: Nijhoff.
Paterson, William (1789) *Narrative of Four Journeys into the Country of the Hottentots, and Caffraria, 1777–79*. London: n.p.
Percival, Robert (1804) *An Account of the Cape of Good Hope*. London: n.p.
Philip, John (1828) *Researches in South Africa*. 2 vols. London.
Raven-Hart, R. (1967) *Before Van Riebeeck: Callers at South Africa from 1488 to 1652*. Cape Town: Struik.
——— (1971) *Cape Good Hope 1652–1702: The First Fifty Years of Dutch Colonisation as Seen by Callers*. 2 vols. Cape Town: Balkema.
Rose, Cowper (1829) *Four Years in Southern Africa*. London: n.p.
Rousseau, Jean-Jacques (1913) *The Social Contract and Discourses*. Trans. G.D.H. Cole. London: Dent.
Sahlins, Marshall (1974) *Stone Age Economics*. London: Tavistock.
Schapera, Isaac (ed. and trans.) (1933) *The Early Cape Hottentots*. Cape Town: Van Riebeeck Society.
Slotkin, J.S. (1965) *Readings in Early Anthropology*. London: Methuen.
Sparrman, Anders (1975) *A Voyage to the Cape of Good Hope . . . 1772–76*. Ed. V.S. Forbes. Trans. Revd J. and I. Rudner. Cape Town: Van Riebeeck Society.
Stavorinus, J.C. (1796) *Reise nach dem Vorgebürge der guten Hoffnung, Java und Bengalen in den Jahren 1768 bis 1771*. Trans. Lueder. Berlin: n.p.
Valentijn, François (1973) *Description of the Cape of Good Hope with the Matters Concerning It*. vol. 2. Ed. E.H. Raidt. Trans. R. Raven-Hart. Cape Town: Van Riebeeck Society.

Van Riebeeck, Jan (1952) *Journals*. Ed. H.B. Thom. Trans. W.P.L. van Zyl *et al.* 3 vols. Cape Town: Balkema.

Vontobel, Klara (1946) *Das Arbeitsethos des deutchen Protestantismus*. Bern: Francke.

Weber, Max (1976) *The Protestant Ethic and the Spirit of Capitalism*. Trans. Talcott Parsons. London: Allen & Unwin.

Wilcocks, R.W. (ed.) (1932) *Carnegie Commission of Investigation: The Poor White Problem in South Africa: A Report*, vol. 2, *The Poor White*. Stellenbosch: Pro-Ecclesia.

6

Punishing violence, sentencing crime

Randall McGowen

Truman Capote's *In Cold Blood* provides a vivid description of the consequences of the murder of the Clutter family for a Kansas community.

> At the time not a soul sleeping in Holcomb heard them – four shotgun blasts that, all told, ended six human lives. But afterward the townspeople, theretofore sufficiently unfearful of each other to seldom trouble to lock their doors, found fantasy re-creating them over and again – those somber explosions that stimulated fires of mistrust in the glare of which many old neighbors viewed each other strangely, and as strangers.

In this passage Capote captures much of the power of violence to shape our lives. The problem of crime for us is framed primarily by the issue of violent offenses, and we read with alarm the criminal statistics in order to discover in them some message about the condition of our civilization.

Capote's language provides insight into another dimension of the issue of violence, not just its ability to spread fear, but its power to produce disorientation. The world is made to seem strange and uncertain. The violent act sets the perpetrator outside of society, not just morally but beyond our rational comprehension as well. Violence has become the domain of the other. Such acts are antithetical to the way in which we imagine ourselves behaving. We like to believe that we cannot understand what makes one person harm another. Yet one consequence of the puzzle over the violence criminals do becomes an uncertainty about ourselves and our neighbors. The violent act comes to define a character as different from us, as criminal. This person appears to be outside of human community, perhaps less than human. But this boundary, seemingly so secure, begins to erode when we become aware of a neighbor's violence, or that of the police. We draw away from these offenders but remain uneasily aware that the violence we thought we had excluded from the community has found its way back into our midst. At such moments we turn away or deny the characterization of the act because otherwise our very identity seems in doubt.

This evaluation of violence has not always prevailed. There was a time when violence was not considered the central ingredient of crime, a time when the state was legitimated by its violence, a time when violent acts were attributed to God's righteousness. Our construction of and preoccupation with violence and its consequences are of recent origin. In this essay I want to explore a chapter in the history of violence, a moment when the concern with violence came to occupy a central place in the elaboration of the criminal law. During the debates over the criminal law in England between 1808 and 1835 the advocates of reform came increasingly to insist upon a distinction between violent and non-violent offenses as a principle upon which to ground the criminal law. This debate was in no sense limited to narrow technicalities; as we will see, it involved the most fundamental issues of social life. An author writing in the *Edinburgh Review* in 1831 demonstrated this point when he complained that existing principles of punishment were defective,

> that no proportion is kept among crimes of different degrees of enormity, and an inducement is thus held out to commit the worse offenses; that the feelings of mankind are apt to run against the punishment, and thus be turned in favor of the offense; and that the frequent spectacle of blood, tends itself to harden the hearts, and corrupt the nature, of the people – thus fitting them for the worst of crimes. (Anon 1831: 402)

In this passage we can observe many of the elements of the modern attitude towards violence being laid down. The author argued that violence was so disturbing that it could delegitimate the authority that employed it. The "spectacle of blood" demonstrated a disregard for physical suffering, but this in turn revealed a frightening state of mind, one that was both individual and collective, the source of atrocious acts. While they were concerned with the pains inflicted upon the body, the ultimate goal of the reformers' program was the mind or soul of the public. In rejecting violence they hoped to promote the ever greater identification of the individual within a collective subject named "the people."

As the title of this essay is meant to suggest, what I want to explore is the way in which a desire to punish violence led to a rewriting of the criminal law. Traditional histories of this episode have accepted the reformers' own account of their motives, but even more importantly have operated within the distinctions the reforms themselves accomplished. In other words, historians have accepted the logic of a criminal code centered upon a concern with individual violence. But a closer reading of the debates reveals that this obvious and taken-for-granted understanding obscures a much more complex and profound story. The theme of violence played through the debates over the criminal laws in a variety of different ways. Without seeking to establish the origin of the reform position, we can none the less reconstruct the various ways violence figured in the debates and examine that social reality which the new discursive regime helped to produce. Although the debates seemed to concern the violence employed by the state and the rejection of that violence in the name of humanity, in fact the controversy served to redefine the conception of civilization, to impose new standards of behavior, as well as to transform the image of the state and the understanding of its activities. Most importantly the preoccupation with violence provided a new code for constructing the individual. On the one hand the attack upon the gallows proclaimed the dignity and value of human life. On the other it suggested the vulnerability of this life to assaults from those at odds with society. It thus increased dependence upon a state that protected one from violence. Within the discourse on violence the reformers elaborated an ideal of social union founded upon subjective feeling. While they claimed that the individual was their primary concern, they more often expressed their hopes and fears in terms of society as a whole. These sympathetic bonds were expected to produce a social order far more smoothly regulated than that of any preceding civilization.

Such a radical transformation was produced by the operation of a seemingly simple distinction: the reformers argued that the law should proportion punishment according to the severity of the offense. This calculus, in turn, rested upon the notion that the best measure of severity lay in the amount of personal violence involved in the offense. The elaboration of this distinction worked to transform the representations of both crime and punishment. It generated a whole series of effects, testifying to new values and goals for society, supporting new social arrangements, and even offering new identities for social groups. The advance of civilization, according to the reformers, was marked by a growth in the concern for the physical wellbeing of the individual. Sympathy, mildness, and benevolence were thought to govern social relations in a progressive state. Those who transgressed the legal boundaries of society seemed a threat that disrupted these relations and contained the seeds of violence. The task of justice, in the eyes of the reformers, was to highlight this danger; the fear thus produced became the central justification for penal measures. Justice protected the peaceful from the violent, who were both outside the civilized community and also represented a regression to the morality of an earlier age. The state acted discreetly but forcefully to protect individual citizens from the perpetrators of violence, employing a minimum of coercion and pain, and with the good of the offender ever a part of its goal. One important consequence of reform was that the actions taken by the state existed in the shade, obscured because of the anxious light directed at the fleeting figures of the dangerous criminals and their violence.

The rulers of English society in the early modern period were, by contrast, unembarrassed by the violence of their punishments. Hanging, branding, flogging, the pillory, drawing and quartering, burning alive – all were accepted as legitimate forms of punishment. In the ceremonies that displayed power and justice the physical suffering of an offender figured prominently. The image of death brooded over every phase of the judicial proceedings. Yet the authorities who made such careful regulations for the rituals surrounding executions seemed, according to later authors, less attentive when they jumbled together a wide variety of different capital offenses in the criminal code. The eighteenth century saw a proliferation of capital offenses ranging from violent acts to instances of forgery and petty thefts. A majority of those sentenced to death had committed no more than crimes against property. Thus although violence was the central element in the punishment of crime, death, the most extreme form of violence, was not imposed on any strict scale that linked it to the violence of the offense. To the reformers, the violence of justice seemed both excessive and often indiscriminate.

Yet we misunderstand early modern society when we describe it in these terms. The goal of punishment was not simply to deter by imposing the most terrifying form of physical suffering. The violence of punishment was a language employed by authority to write the message of justice. The afflictions suffered by the body of the condemned were a way of representing lessons. A hanging, for instance, was meant to be a frightening example but also an eloquent one, speaking not only the death of the individual but also the divine sources of secular power and the horrifying punishments awaiting him in the next world. The broken body of the condemned represented the restored order of the body politic. Punishments were excessive because they bespoke the disproportion between an individual and the majesty of God and the law. The order and hierarchy that alone made human life possible could be threatened by recalcitrant human nature. God employed the indescribable sufferings of hell as a warning to instruct humanity, and these afflictions were justified not by the severity of the offense but by the character of the infinite authority that had been offended. Human governments relied upon the fear of God, and human punishments were intended to represent the sacred lesson. The violence of punishment pointed to the power and sanctity of authority (McGowen 1987).

Defenders of the existing criminal laws were not unaware of the charges made against prevailing punishments, though by the second decade of the century their complaints against the reformers suggest a feeling of impotence. In 1830 the Tory Charles Wetherell complained that "the opinion that punishment by death was as impolitic as it was cruel, in offences against property, was daily making way, was creeping over the face of public opinion." He warned that severity was needed to overawe an unruly populace, and that force lent dignity to those who used it. The reformers were to be condemned because they were trying "to disarm Government of its powers" (*PD* 1830: 2, 25, 64–5). Faced with calls to make a sharp distinction between offenses with and without violence, a Lord Chancellor protested that he "was not prepared to draw the line so distinctly" (*PD* 1831: 3, 6, 1,177). To judges such as Eldon and Ellenborough public executions testified to the will to rule of those in authority. Criminal justice was theater, and severe punishments marked the seriousness of the occasion. They dignified the state, and they imparted a reflected glow to those who exercised a discretion that could regulate punishment (Hay 1975; McGowen 1983). The severity of the criminal law found other justifications as well. It was held to be a necessary feature of a society where subjects had many liberties and where police measures were limited. Solemn examples were also deemed important for those offenses, such as cattle and sheep stealing, where property was exposed and detection difficult. Tories argued that forgers should be hanged because their very existence placed a rich commercial society at risk. "It had always been deemed necessary," commented a Member of Parliament, "to punish those offences that could be committed with great facility, with more severity than those of a contrary description" (*PD* 1808: 1, 11, 994). The seemingly extravagant character of punishment found many sanctions; perhaps the most surprising omission from this list was any mention of violence involved in the offense itself. In other words the debate over the criminal laws centered upon a distinction that scarcely existed in and certainly did not define the character of the earlier criminal code.

The advocates of reform displayed little patience with the arguments made in defense of the existing criminal code. They argued that in prescribing death for so many offenses, often of a trivial nature, the law demonstrated a disregard for human life. In one debate in 1819 T.F. Buxton proclaimed to his opponents that "your law is vicious in principle," while William Wilberforce called it "that code of blood." James Mackintosh, a Whig with an extensive philosophical background, sought to prove that these laws were "the relics of barbarous times": they were "disgraceful to the character of a thinking and enlightened people" (*PD* 1819: 1, 39, 807, 829, 791). By 1832 even a Tory such as Lord Wynford found himself acknowledging the crucial distinction upon which the reform cases came to rest. He agreed that "in none but extreme cases" should the punishment of death be inflicted. Hanging was just, he concluded, where life was "wantonly" taken away, or where, in cases of burglary or robbery, great personal violence was used. "Except in cases of this description, he considered the infliction of death as bloody, barbarous, and unnecessary" (*PD* 1832: 3, 13, 991–2). Wynford made it clear that the violence that had so seldom been the focus of attention, the violence of the offense itself, had now become the central issue under consideration. The problem with the violence of the execution, in the opinion of the reformers, was that it obscured and confused the guilt of the violent offender. Physical suffering was not a desirable element of punishment. Lord Holland hastened to establish a distinguished lineage for this distinction: he thought that

> the common feelings of mankind, the maxims of religion and philosophy, the authority of the most eminent men, and the practice of the most civilized nations as well as of our ancient law, are generally averse to

> punishing by death any crimes in the perpetration of which no violence is used or intended. (*PD* 1830: 2, 25, 1,163)

Another Member of Parliament said simply: "The penalty of death ought not to be attached to crimes which were not attended by violence and terror" (*PD* 1833: 3, 15, 1,158). The reformers focused all of their attention and anger upon the violence of the laws, but they did so without any desire to diminish the concern felt for the violence committed by criminals. They demanded reform as part of a campaign to circumscribe and eliminate violence from the life of their society. In the future only criminals would be responsible for the existence of violence in English civilization.

The most familiar and powerful convention of the reform position was the use made of history to valorize the distinction between kinds of offenses and to justify the amelioration of the laws. Violence was invariably associated with an earlier, more primitive level of society. Indeed this violence was distinguished as the central moral feature of these civilizations. As one legal author, William Roscoe, wrote: "Wherever we turn our eyes on past ages, we may observe the same insensibility to the suffering of others, and the same propensity to, not to say eagerness for, the shedding of human blood" (Roscoe 1819: 6). A frequent pamphleteer in the reform cause, Basil Montagu, developed the narrative in great detail. "In a savage state," he argued, "the prisoners are tortured by every pain which the mind of man, ingenious in cruelty can invent." He listed in exquisite detail the catalog of inflictions developed in "past ages" and applied to "his fellow-creatures." Humans tortured each other "with unrelenting ingenuity," "by tearing limbs asunder; by racking; by breaking upon the wheel," and in a hundred other ways. One source of dismay for him was that "such atrocities" were "not confined to the infancy of society" (Montagu 1821: 5–6). He found a natural analogy and explanation for this history in the idea of the development from childhood to adulthood. "In the infancy of his reason savage man, a slave to turbulent passion, delights in revenge." "In the infancy of society" this same "ferocious spirit" was transferred to the public. "Severity thus gains an ascendency in all countries; and being once established, it is, from our tenacity in retaining existing customs, the work of centuries to extirpate the effects of passion by the dictates of reason." He believed that the advance of civilization was slow, but at the same time he thought that it was "ultimately certain and irresistible." Knowledge and humanity advanced together. "Torture is no longer used in Europe." "The wheel is abolished," he added confidently, "and there is now a conflict with respect to the expediency of sanguinary punishments for crimes without violence" (Montagu 1827: 19). Montagu argued that past punishments were simply the product of passion and the thirst for blood and excitement, while the advance of reason, compassion for suffering, and a higher regard for the individual were displayed in the demands for milder punishments. Punishment provided a key to the reading of an era and to deciphering the quality of its inner life. Past governments celebrated pain in public displays and offered a more benign and pleasing image. The presence or absence of violence became the standard by which a government and a civilization were judged.

Exactly because external signs indicated internal states the reformers were made uneasy by the condition of their criminal law. The gallows cast a gloomy shadow over the happy descriptions of their own age. It raised unsettling questions about the accuracy of their self-identification as humane and rational. How would citizens of other European nations see them? "If a foreigner," suggested Mackintosh, "were to form his estimate of the people of England from a consideration of their penal code, he would undoubtedly conclude that they were a nation of barbarians" (*PD* 1823: 2, 9, 397). Even the Tory Robert Peel was forced to concede that "the multiplication of this threat in the laws of England has bought on them and the nation a character of harshness and cruelty which evidence of a mild administration of them will not entirely remove" (*PD* 1830: 2, 23, 1,181). The reformers wished to disprove the accusation that they were insensitive to human suffering and tolerant of violence. They hoped to establish just the reverse. One legal author, with a touch of defensiveness, made much of the contrast between the present and England's "barbaric ancestors." "With all our vices," he noted, "their savage, cruel, and vindictive habits, will endure no comparison with our humanity and philanthropy" (Putt 1830: 173). The reformers established their claim that the treatment of the body was simply a test that revealed an inner disposition. Acts were most important for what they displayed of subjective consciousness. Thus the task of reform was from the first involved with the regulation and projection of images, in order to control the mental and moral life of the country.

What was it about violence that made it so destructive? According to one author "the effects of cruelty and severity" were "to strike out of man all feelings of sensibility." They tended to blunt the mind "and render all within dark" (Anon 1833: 232). Violence stood opposed to a social principle that the reformers desired to promote above all others, that of sympathy. This ideal was implied in every adjective used to characterize the evils of the past and the virtues of the present. Sympathy suggested a society in which individuals discovered in the depths of their souls a concern for and identification with others in society. People willingly regulated their own conduct so as to produce a minimum of uneasiness or misunderstanding. This process demanded of the individual ever greater openness and sensitivity. A tenderness of mind produced mild behavior. In a civilized society the social affections reigned. Sympathetic union was the reverse of the condition of social relations in ruder ages. In those societies, as we have seen the reformers argue, individuals possessed a savage indifference to the feelings of others; they were hard-hearted and cruel. They gave vent to their own selfish passions and acted with an irrational disregard for consequences.

Whatever other noble qualities they might possess, these virtues were negated by their unfeeling natures. Disorder and violence sprang from a lack of mutuality and concern for others. An act of violence was a symptom of a breakdown of social sympathy on the part of the perpetrator and marked his regression to an earlier moral state. In the reformers' eyes the act itself was less frightening than what it indicated about the condition of the offender. They were alarmed by the evidence of social fragility. They proposed not a counter-terror, but a series of measures aimed at the criminal that would restore a sympathy seemingly more perfect than any mere code of laws could create (McGowen 1986).

The reformers admitted that another feature of the prevailing criminal code that excited their anger and condemnation was the frequent executions for property offenses. If on the one hand the gallows suggested that they valued life too little, the code implied that they regarded property too highly. Once again the implication was that they did not pay enough attention to the quality of social relations. "Kill your father, or catch a rabbit in a warren," exclaimed Buxton, "the penalty is the same!" (*PD* 1821: 2, 5, 905). Mackintosh voiced the same complaint when he said that "the feelings of the present age did not allow of the barbarous aggravations of death practiced by our ancestors . . . [when] we hanged alike the sheep-stealer and the parricide" (*PD* 1830: 2, 24, 1,038). "Is it fit," asked still another author in 1816, "that for crimes, which merely abridge the comforts of those around them, that men should be condemned to death?" (Polidari 1816: 288). The confusion of the old criminal code, they feared, might not be accidental. It was the imposition of death for so many property offenses that marked the code as sanguinary and demonstrated how the government undervalued humanity. "Woe to the government," wrote James Mill, "whose laws . . . tend to level its distinctions and obliterate its precepts!" He claimed that "human nature recoils at the thought of spilling blood – yet a law positively commanding murder would be scarcely less vicious in principle and tendency, than one which declares, that murder is not a more heinous offense than horse-stealing" (Mill 1811: 155). The law undoubtedly existed to protect property, but its true purpose, the reformers argued, as measured by those offenses it should take most seriously, was to protect life.

Such a principle seemed suited to an age that was increasingly democratic as well as civilized. Great inequalities of wealth existed but everyone had an interest in preserving his or her own life. The Duke of Sussex, in presenting a petition from London grand jurors, spoke of the need "to disconnect the punishment of death from crimes affecting merely the rights of property, but unaccompanied with violence or bloodshed." The petition itself argued that a wise government ought "not teach cruelty to the people." The current law did so. The jurors concluded by calling for the amendment of the law "by drawing a distinction between the simple invasion of the rights of property, and crimes of violence and blood" (*PD* 1831: 3, 6, 1,172–5). The poor were supposed to learn that the law was primarily concerned with protecting what they shared with the rich, their lives. They were also taught to respect the lives of the wealthy as well as their own. Even the criminal should learn to distinguish between theft and murder. In this way a justice still overwhelmingly occupied with the prosecution of offenses against property sheltered within a code whose rhetoric and schedule of punishments proclaimed that it valued something else more highly.

If the law acted to unite rich and poor in an appreciation of how it operated to protect their lives, it also served to mark off in a particularly striking way those who committed a violent offense. The emphasis upon violence heightened the fear of violence, or, as one Member of Parliament expressed the thought, the moments when a citizen experienced "bodily fear of his life" (*PD* 1833: 3, 17, 165). Sympathy extended only to those who refrained from physical threats; the harshest language of the debates was reserved for those who perpetrated atrocious offenses. The very language of barbarism that was used to characterize a law in need of reform was also employed to describe those offenders. "Reason certainly it seems," wrote an advocate of reform, "that an equal measure of punishment should be inflicted upon such an execrable wretch who has shown himself at war with his fellow creature, and torn asunder the bonds of human society" (Putt 1830: 167). The virtue of a reformed code would be the way it would isolate and dramatize violent offenses. It would unite society in hostility to the actions of this class of criminal. No one spoke against preserving severe punishments for these offenders. "Crimes of violence," cautioned a solicitor general, "must continue to be under the painful necessity of visiting with death. For murder, for burglary, for arson, for all crimes which threatened life, it seemed to him that the punishment of death accorded with the natural feelings of man" (*PD* 1833: 3, 17, 164).

While the reformers believed that the law might be employed to secure the closer union of the members of society, they feared that the existing code produced a different result. "The laws so constructed," wrote one author, seemed "to divide society"; they "set one part against the other, tending to destroy the bond of reciprocity between the upper and lower classes" (Anon 1833: 14). If the reformers were eager to differentiate their own age from past societies, they were worried at the appearance of that opposition between civilized and savage within their society. One evil of the gallows was that it lowered the threshold and seemingly legitimated the recourse to violence. This danger was greatest among the lower orders for whom the result of a hanging was that "the number of ruffians" was increased "by the removal of a few whose mantle descends on their successors, with a double portion of their spirit" (Anon 1814: 137). The reformers held that the confusion of categories in the criminal law was not simply an intellectual matter, but muddled the understanding most people had of the social world. It confused the normal lines of sympathy and antipathy in society. The commissioners writing on the criminal law in 1836 found it necessary to remind their readers of this distinction: "The efficiency of penal laws is to be judged, not as they are likely to operate on the greater mass of mankind, who are restrained from the commission of crimes by a sense of religion and morality, respect for public opinion, love of their families and connections." The damage produced by the existing code was that it created a sympathy for those who did not deserve it and yet aroused no reciprocating feeling of gratitude in the criminal. Here was "a class of society whose habits have excluded them from any honest mode of life, whilst a course of vicious sensual indulgence, and an aversion to labor, have increased in a great degree the temptation under which they lie to commit crimes." The capital code encouraged them to resort to the most extreme and violent actions (*PP* 1836: 20).

While expressing the desire to unite all levels of society in a common appreciation of the sacredness of life and in opposition to the violent offender, another fear appeared in their language. Public inflictions of pain tended to undermine all social feeling and civilized standards. But the danger was greatest for the mass of the people who had made the least progress towards civilization. "The cruelty of our laws," wrote one author, tended "to perfect the brutalization of that class of society who are engaged in crime, and also all those who, from this want of education and position in life, are the most liable to fall into it" (Anon 1833: 201). Among the lower orders many forms of violence still found acceptance, such as bear-baiting or cockfighting. These were the people most likely to attend an execution and to make it the occasion for a holiday. With this class in mind, one author observed that "public executions inspire spectators with a savage taste for slaughter. The thirst of blood which causes such wide desolation is consecrated by the example of the laws." He feared the consequences were predictable: "The most tremendous and chilling ideas become familiar and unimpressive. Murder produces murder, death generates death; and the contagion of scaffolds is diffused through the life-blood of nations" (Anon 1814: 135). An execution could result in the very identification of crowd and condemned that the reformers were so concerned to prevent. William Roscoe, a legal authority, was puzzled that legislators and judges had never made the obvious deduction; he wondered if "it never occurred" to the advocate of death that "by exhibiting frequent and revolting spectacles of inhumanity and bloodshed, he has counteracted his own object, and weakened in the public mind that natural reluctance to the shedding of human blood, which is one of the great safeguards of civil society?" He was convinced that "a spirit of cruelty and persecution is awakened, which is sometimes carried to such an extreme, as perhaps to be scarcely less criminal than the offense which it is intended to avenge."

> In order to demonstrate to a people, that they ought not to be cruel, he sets an example of cruelty; and, in order to deter them from putting each

other to death, he puts them to death himself; and that frequently by such acts of inhuman atrocity and savage barbarity, as the most ferocious criminal was never known to commit.

Was it any wonder that severe penalties encouraged criminality, and increased "both in frequency and enormity" the incidence of crime, heightening the sense of personal danger felt by all members of society (Roscoe 1819: 8. 17).

The reformers took great care in analyzing what they took to be the psychological consequences of harsh punishments. These inflictions, they believed, tended to confuse ideas of right and wrong, and disorder social feeling by severing the sympathy with those above while fostering it with those below. The British criminal law, Buxton warned, "in its origin and source, was peculiarly merciful and tender of human life." "But, upon this system which, resting upon public cooperation, requires the public sympathy, we have unhappily engrafted another system, which does violence to the feelings of the nation" (*PD* 1821: 2, 5, 925). Public displays of suffering were uncontrollable in their effects; they often did more harm to the innocent than to the lawbreaker. The operation of justice even tended to blur the line distinguishing the violent offender and the actions of lawful authority. "When mankind beheld the life of a fellow-creature," announced Mackintosh to Parliament, "sacrificed for a petty theft, a trifling injury or fraud, their feelings at once revolted, they sympathized with the sufferer in his dying moments, and, ascribing his punishment to the effects of superior power alone, they too often inwardly loaded both laws and judges with execrations" (*PD* 1822: 2, 7, 794). The reformers expressed alarm at the unpopularity of the law and at the way it reinforced the existing divisions in the community. The error seemed to spring from mistaken opinions about the relationship of violence and property. In the words of one author, "there should be a marked and acknowledged difference in the punishment of crimes, or there must be a loss of all moral distinction, or of all respect for the laws that systematically held it in complete scorn" (Anon 1821: 347).

While the danger seemed so great and threatening, the remedy to this desperate situation seemed deceptively simple. Buxton expressed the solution when he argued that it was "a principle which nature has implanted in the heart of man, and which no human law can eradicate – that, as there are distinctions in guilt, so there should be distinctions in the penalties annexed to it"(*PD* 1819: 1, 39, 810). In 1830 Mackintosh appealed to the French model, where of 110 executions over a period of years, 90 were for murder, to demonstrate the correct balance. In France the law operated on the basis of a principle consistent with the advance of civilized standards: "This indicated a just execution of the law – an application of the last punishment to a class of offenses that naturally required it, and to which it ought to be confined." The degrees of punishment, he contended, were to be found in the amount of violence in the offense. The entire scale of punishments would then be anchored in a just appreciation of the most severe of illegalities. For Mackintosh the beauty of such a law lay in its transparency and naturalness; it would not draw attention to itself, but rather act as a carrier of collective sentiment. It would become a powerful motive and force sustaining order. "The acts for which the punishment of death should be applied, should not only be in the highest degree dangerous," he concluded, "but attended with circumstances of violence and blood, having a deep impression on the mind, and reviving indignation at the offender on the recollection of his crime" (*PD* 1830: 2, 24, 1,038–41). By directing attention to crimes of violence, the reformers believed that they could exercise control over public sentiments. Such was the thrust of G.B. Mainwaring's testimony before a parliamentary committee when he said that "punishments, to be universally operative, must be such as are in unison with the common feelings of humanity. . . . Against crimes of an atrocious nature, the public mind will rise to the highest degree of punishment which the law can denounce" (*PP* 1819: 8 (586) 121).

Much of the force of the reformers' solution derived from the argument that the distinction they proposed was natural. This contention both justified and ensured the success of their measures. Yet every example they offered suggests how uncertain they felt about this naturalness, and how little they could rely upon it. The vehemence of their arguments emphasizes that the distinction, far from being natural, was one they had created and would struggle to maintain. The history of progress they offered to justify their efforts raised cause for concern. Mildness and sympathy in human relations were of too recent origin to be deemed natural. Indeed the desire to make an artificial distinction natural provided the impetus behind much of what they said and wrote. They returned frequently to the topic of the necessity of shaping and moulding public character and of the task of fashioning a society in which sympathy governed human conduct. Mackintosh believed that "the reduction of capital punishment would tend to the improvement of civilization" (*PD* 1822: 2, 7, 797). "The old, obvious, and common argument in support of the revision of the capital denunciations," he suggested upon another occasion, "was founded on the feelings of common humanity, and on that tender regard to life, which it was at all times so desirable to inculcate; and which it became the special duty of the lawgiver to teach by his commandments, and encourage by his example" (*PD* 1820: 2, 1, 228). The distinction between violent and non-violent offenses needed to be taught time and time again. The law deferred to sentiment, acknowledging its power, even as the reformers employed the law to shape and sway opinion. "Knowing that the sentiments of society have a powerful effect in enforcing laws," wrote Montagu, the wise legislator

> will make the laws react powerfully on the sentiments of society. He will not permit them either to fall short, or disappoint the just indignation and

> moral sense of the community, or exceed and outrage it. He will satisfy the sense of justice in the public mind, and strengthen the opinion of the community. He will not, on the one side, suffer the arm of justice to be mocked and baffled by the impunity of offenses, nor to be unnerved by thwarting and prevaricating with the common sentiments of mankind. (Montagu 1821: 29–30)

On the face of it the reformers seemed to lack confidence in the authority or the ability of the state to wield violence. But a more careful reading of their arguments suggests that they had discovered a more efficient application of power to achieve their ends.

Justice as displayed at the gallows was now judged to be too blunt, excessive, and unwieldy. If the law needed to show restraint as well as subtlety this was because its goals were now more ambitious. The reformers found terror no longer an adequate defense for society. The law required more circumspection in its operation. The commissioners on the criminal law in 1836 concluded that the "want of discrimination in the denunciation of punishment is impolitic." "It may be hoped," they added, "that the law, in the annexation of severe and appropriate penalties to heinous offences, may operate not only on the fears, but also on the consciences and feelings of mankind, and serve to impress their minds with just conceptions of the comparative enormity of crimes" (*PP* 1836: 36 (343) 28). The radical William Ewart used almost the same words when he called upon Parliament to "educate your people." "Be not content," he warned the members, "with giving them a horror of punishment – give them a horror of crime also" (*PD* 1832: 3, 11, 951). The task of the justice was no longer accomplished by what punishment enacted, but also by what it made a display of not doing, of the pain it refused to inflict. This message was made all the sharper because the violence it declined to employ was exactly what the most dangerous offenders were guilty of using. The traditional criminal code had operated on the basis of analogies to the crime and to future punishments; the reformed code sought to set up a contrast between the intentions of the offender and the state. The goal more specifically was to make the horror of crime exceed the horror of punishment; this was the only way to satisfy Lord Holland's injunction that "the compassion likely to be produced by the punishment should not exceed the indignation generally excited by the perpetration of the crime" (*PD* 1830: 2, 25, 1,164).

The reformers hoped that violence would mark the criminal as existing outside of society and presenting a brutal threat to its physical wellbeing. The violent offender served to define a community within which peace and order prevailed. A heightened awareness of such a class could be of immense use to a society. Such an idea would discourage honest citizens from resorting to violent means since all violence was in some sense criminalized with the reformers' scheme. Mackintosh said he would "make the penal laws of his country the representative of the public conscience, and would array it with all the awful authority to be derived from such a consideration. He would make it the fruit of moral sentiment, in order to render it the school of public discipline. He would array the feelings of all good men against the dangerous criminal, and would place him in that moral solitude where all the members of society should be opposed to him'' (*PD* 1823: 2, 9, 418–19). The reformers aimed to use the law to tighten the definition of this alien group, to exile them more effectively from the rest of society, and to affix to them more securely the title of enemies of humanity. For all the sympathy they seemed to address to criminals and for all their desire to mitigate the punishments they suffered, the reformers' proposal also aimed to increase social distance and to darken the colors in which criminals were painted.

One consequence of this transformation in the attitude towards violence, and not the least important, was that the representation of those measures taken by the state to deal with criminals was profoundly altered. The old regime was characterized as using methods that were mistaken and ineffective; as one Member of Parliament observed, ''brutal treatment, hanging, and gibbeting were neither the most economical nor the most efficacious, as they were certainly neither the most humane nor the most enlightened modes of punishing crime or reforming society'' (*PD* 1833: 3, 17, 162). The rhetoric of the execution was too crude. The reformers, on the other hand, hoped to turn the difference between a hanging and the prison to their advantage. While death attracted public sympathy for the condemned, James Harmer thought that ''any thing short of death the public appears to be satisfied with, however severe; there is no sympathy for a delinquent if his life is spared, which is a great thing to accomplish; for, in my opinion, no punishment can be effective which excites a feeling of compassion for the sufferer'' (*PP* 1836: 36 (343) 81). The public was taught to identify with the refusal to use violence, with the benign measures taken to promote the reform of the prisoner, and with the goal of making the offender more like oneself. Punishments should pacify the criminal and public at large. ''As even savage beasts,'' wrote a legal author, ''such as bears and lions, can be made tractable and docile, so the most savage amongst us, by a judicious imprisonment, accompanied with reflection and religion might be ameliorated and reformed'' (Putt 1830: 173). Government, turning away from violence, instructed people in gentleness. The reform narrative did not refrain from speaking of death; it spoke of it more volubly, in order to emphasize its own title to humanity.

Just as the displays of violence at executions in the eighteenth century had everything to do with a dramatic portrayal of state power, so reform was preoccupied with the representation of government. The distinction between violent and non-violent offenses operated to focus public attention upon those acts that threatened their lives, as if such acts constituted the majority of criminal offenses or the main challenge presented by the offender. The state in the shape of the police and prison now appeared as the benign protector of the public or the anonymous but benevolent administrator of a confined population. As one author wrote:

> The question would no longer be, whether stripes and bloodshed can prevail against guilt and ignorance, but whether sympathy, prudence, and compassion, have lost their influence on the moral feelings of mankind. If we could impress upon the mind of the delinquent, an idea that the efforts we are making are really intended for his welfare, our object would in great degree be accomplished. There is no human being so stupid, or so wicked, as not to concur, to the utmost of his power, in measures evidently calculated to relieve him from misery. (Roscoe 1819: 9)

The goal of imprisonment was to reform the offender so that he or she could return to society. Whereas the gallows alienated "the best feelings of human nature," the prison, according to Thomas Wrightson, conciliated those feelings. It displayed "real kindness towards the prisoner, inasmuch as it presents the means best adapted to effect his reformation – which is the truest humanity" (Wrightson 1833: 29). Punishment operated on those who by their actions indicated that they were unfitted for civilized life by employing the smallest measure of suffering possible. Criminality was evidence of a hardened disposition and deadened feelings. The prison sought to arouse their affections, the emotions one author called "the keys to the mind" (Anon 1833: 232). The criminal had withdrawn from the community; the goal of reformation was to produce in the offender the desire for social sympathy and a willingness to regulate his or her conduct so as to achieve it. Once more in harmony with the collective subjectivity of a civilized society the reformed individual was ready to return to the world.

Although the target of the prison was the transformation of the individual, the reformers remained conscious of the need to control the image of justice. "Instead, therefore, of connecting the ideas of *crime* and *punishment*," wrote Roscoe, "we ought rather to place together the ideas of *crime* and *reformation*; considering *punishment* as only *one of the modes* for effecting such reformation, the extent of which must always be proportioned to the necessity of the cause" (Roscoe 1819: 10). Capital punishment remained in the criminal code, but it served to reinforce the general scheme; one speaker suggested that it should be "confined to murder, and to such instances of personal violence as might lead to the loss of reason or of bodily health" (*PD* 1830: 2, 25, 591). In such cases the state did not act of its own will, but as a surrogate for the one who suffered and so as to symbolize that it acknowledged the unique importance of each individual. Once punishment had celebrated the terrifying power of the law; now it was the terror of individual violence that became the focus of anxious attention.

We can read the debate over the criminal laws in early-nineteenth-century England from several different perspectives. In part a society that recognized the force and legitimacy of public opinion found it necessary to convince people that the laws "ought to be looked up to as an equal safeguard and an equal source of relief and satisfaction for all" (Anon 1815: 225–6). Others argued that reform of the law was demanded by the progress of civilization. Ewart claimed that "a mild code of law was really a proof of a great and good country, because, in all countries where the law was mild, the inhabitants were also mild and civilized" (*PD* 1833: 3, 17, 172). Gentle punishments taught the people lessons of civilized conduct, while a code that punished personal violence most severely displayed a high regard for human life. According to Montagu, in a "civilized state" "all encouragement is given to kindly affections, and dissocial passions are softened, if not subdued by habitual submission to legal authority" (Montagu 1818: 103). In all of their accounts the reformers portrayed themselves as passive agents of forces and influences beyond their control. At the same time they championed the extension of these controlling movements to the rest of society. The reformers demanded a changed criminal code that both ended displays of public violence and punished the violence that criminals committed with greater emphasis. The argument of this essay is that the distinction they insisted upon had wideranging if subtle consequences. The discourse upon violence produced a new way of reading society. It generated a different set of responses to crime along with a different constitution of criminality from that of the earlier code. These images of crime demanded new remedies that in turn projected a different fact of government. The success of this project can be measured by the fact that while we are aware of the violence committed by the prison, none the less the weight of our fears is summed up in the violence that we attribute to criminals and from which the prison is still believed to preserve us.

References

Anon (1814) "Strictures on the Right, Expedience, and Indiscriminate Denunciation of Capital Punishment." *Pamphleteer* 3: 115–54.

——— (1815) "Capital Punishment." *Philanthropist* 5: 222–8.

——— (1821) "Capital Punishments." *Edinburgh Review* 35: 314–52.

——— (1831) "Capital Punishment of Forgery." *Edinburgh Review* 52: 398–410.

——— (1833) *Old Bailey Experience: Criminal Jurisprudence and the Actual Working of Our Penal Code*. London: n.p.

Hay, Douglas (1975) "Property, Authority, and the Criminal Law," in Hay *et al.* (eds) *Albion's Fatal Tree*. New York: Pantheon: 17–63.

McGowen, Randall (1983) "The Image of Justice and Reform of the Criminal Law in Early-Nineteenth-Century England." *Buffalo Law Review* 32: 18–125.

——— (1986) "A Powerful Sympathy: Terror, the Prison, and Humanitarian Reform in Early-Nineteenth-Century Britain." *Journal of British Studies* 25: 312–34.

——— (1987) "The Body and Punishment in Eighteenth-Century England." *Journal of Modern History* 59: 651–79.

Mill, James (1811) "On the Penal Laws of England." *Philanthropist* 1: 66–77, 143–56.

Montagu, Basil (1818) *Some Inquiries Respecting the Punishment of Death for Crime Without Violence*. London: n.p.

——— (1821) *Thoughts Upon the Abolition of the Punishment of Death in Cases of Bankruptcy*. London: n.p.

——— (1827) *Reform*. London: n.p.

Parliamentary Debates (*PD*) (London) year: series, volume, column.

Parliamentary Papers (*PP*) (London) (1819) "Report from the Select Committee on the Criminal Law Relating to Capital Punishment." *PP* 1819, 8 (586).

——— (1836) "Second Report from the Royal Commission on Criminal Law." *PP* 1836, 36 (343).

Polidari, John (1816) "On the Punishment of Death." *Pamphleteer* 8: 281–304.

Putt, Charles (1830) *Essay on Civil Policy*. London: n.p.

Roscoe, William (1819) *Observations on Penal Jurisprudence*. London: n.p.

Wrightson, Thomas (1833) *On the Punishment of Death*. London: n.p.

7

Hysteria and the end of carnival

Festivity and bourgeois neurosis

Allon White

> While she was being massaged she told me only that the children's governess had brought her an ethnological atlas and that some of the pictures in it of American Indians dressed up as animals had given her a great shock. "Only think, if they came to life!" (she shuddered). I instructed her not to be frightened of the pictures of the Red Indians but to laugh heartily at them. And this did in fact happen after she had woken up: she looked at the book, asked whether I had seen it, opened it at the page and laughed at the grotesque figures, without a trace of fear and without any strain in her features. (Sigmund Freud, *Studies on Hysteria* (1888–9))

I am fascinated by the carnival debris which spills out of the mouths of those terrified Viennese women in Freud's *Studies on Hysteria*. "Don't you hear the horses stamping in the circus?" Frau Emmy implores Freud at a moment of particularly abject confusion. It is striking how the broken fragments of carnival, terrifying and disconnected, glide through the appalled discourse of the hysteric. Occasionally, as in the extract quoted above, it appears that Freud's therapeutic project was simply the reinflexion of this grotesque material into comic form. When Frau Emmy can at last look at the "grotesque figures" of the American Indians dressed as animals and "laugh without a trace of fear," it is as if Freud had managed a singular restitution, salvaging torn shreds of carnival from their phobic alienation in the bourgeois unconscious by making them once again the object of cathartic laughter. It is, of course, significant that the carnivalesque practice which produced the phobic symptom in Frau Emmy is that of an alien, non-European culture. Not the least significant element in the suppression of the indigenous carnival tradition in Europe was a compensatory plundering of ethnographic material – masks, rituals, symbols – from colonized cultures. In this respect, Joseph Conrad was doing no more than Frau Emmy von N. in placing "savage rites" at the heart of European darkness in the 1890s.

As we know, within a very few years Freud was to abandon the carthartic approach that he used with his early hysterical patients and was no longer interested in trying to precipitate the abreactive rituals in his patients which, as above, might reinflect the grotesque and disgusting into a *comic* form. An excellent, painstaking examination of the *Studies on Hysteria* by Scheff (1979) suggests that Freud's decision to abandon a cathartic approach was premature and based upon a determinedly sexist interpretation of his success as partial failure, particularly in relation to Anna O. Indeed, it is a notable feature of the case histories that it is *the patients themselves* who, in their pastiche appropriations of festive, carnival, religious, and pantomimic gestures, suggest kinds of alleviation to their own suffering. Anna O. is credited by Ernest Jones as being the real discoverer of the cathartic method and Breuer developed and formalized her practical notions in his own early method (Scheff 1979: 28). Freud's gradual move away from abreactive rituals of a cathartic kind towards associative methods of self-consciousness is entirely consonant with his desire to produce a *scientific* psychology. This is because science – and particularly acutely during the late nineteenth century – has always been deeply hostile to ritual. Indeed, science only emerged as an autonomous set of discursive values after a prolonged struggle against ritual. It marks its own identity by the distance it established from "mere superstition," which is science's stigmatizing label for, among other things, a large body of social practices of a therapeutic and arguably useful kind. Scheff has suggested that there is a strongly cognitive bias against ritual and catharsis in recent work in anthropology, psychotherapy, and psychology: "the emphasis has shifted away from the catharsis to a concern for cognitive and symbolic functions" (Scheff 1979: 21). A rereading of hysteria case studies, in relation to both ritualistic and symbolic carnival material, suggests new ways in which the hostility of rational knowledge to ritual behavior might be resolved.

I am struck forcibly, as I reread the *Studies on Hysteria*, by how many images and symbols, which were once the source of vigorous pleasures in European carnivals, have become transformed into the morbid symptoms of private terror. Again and again these patients suffer acute attacks of disgust, literally vomiting out residue of traditional carnival practices. At the same time, the women seem to be reaching out, in their highly stylized gestures and discourses, toward a repertoire of carnival material for both expression and support. They attempt to mediate their terrors by enacting private, made-up carnivals. In the absence of social or communal forms, they attempt to produce their own by pastiche and parody – in an effort to semiotically embody their distress. Once noticed, it becomes apparent that there is another narrative – fragmented and marginalized by the emergent psychoanalytic discourse – but lodged within it. This "other narrative" tells, in hints and scraps of carnival symbolism and parodic festive behavior, of the individual bourgeois subject trying to produce its own private – and therefore doomed – therapeutic rituals from discursive bits of carnival debris. In his general remarks on hysterical attacks, Freud himself even makes this process part of his definition of hysteria, without, however, making anything of it. "When one psychoanalyzes a patient subject to hysterical attacks one soon gains the conviction that these attacks are nothing but phantasies projected and translated into motor activity and *represented in pantomime*" (Freud 1909: 153; my italics). Freud goes on to talk about the distortion which "the pantomimic representation of phantasy" undergoes as a result of censorship. Yet such is the low status of ritual and dramatic representation that Freud never "sees" his own reference to European carnival as anything other than metaphorical. Towards the end of a letter to Fliess, written in 1896, Freud remarks on what Charcot had dubbed the "clownism" phase of hysterical attacks. Freud remarks on "the 'clownism' in boys' hysteria, the imitation of animals and circus scenes . . . a compulsion to repeat dating from their youth (in which they) seek their satisfaction to the accompaniment of the craziest capers, somersaults and grimaces" (Freud 1954: 181–2).

Even though Charcot's typology of hysteria is unreliable and contrived, especially for the photographic representations, the "clownism" was well attested as a symptomatic aspect of hysteria. Freud, in the explanation he offers to Fliess in the letter, refers to the "perversion of the seducers" (the patients themselves) who, he says, "connect up nursery games and sexual scenes." But this connection is arbitrary and unconvincing, especially when seen in relation to the whole range of festive material scattered through the studies on hysteria. There are indeed deep connections between childhood games as found in the lore of schoolchildren and carnivalesque practices of European culture (White 1983), but here Freud's insistence of a purely sexual aetiology and an individualist perspective obscures a fundamental sociohistorical matrix of his narrative. The carnival material in the case studies witnesses an historical repression and return. The repression was the gradual, relentless attack on "the grotesque body" of carnival by emergent bourgeois hegemony from the Renaissance onward. Interestingly, scholars of European popular culture have occasionally wanted to connect up, backwards as it were, Renaissance festive form to Freud's ideas. Thus Barber talks generally of "A saturnalian attitude, assumed by a clear-cut gesture toward liberty, brings mirth, an accession of wanton vitality. In the terms of Freud's analysis of wit, the energy normally occupied in maintaining inhibition is freed for celebration" (Barber 1959: 7). But Barber's reference to Freud here seems to me to be reaching for a validation that confuses the historically complex relation of the discourse of psychoanalysis to festive tradition. The history of the repression of carnival and popular culture has to be related to the victorious emergence of specifically bourgeois practices and languages which reinflected and incorporated that material within an individual framework. In one way or another, Freud's patients can be seen as enacting desperate, private ritualistic fragments salvaged from a festive tradition which, by 1900, had been more or less eliminated. Thus the "highly gifted lady" of the case studies celebrated a whole series of what she called "festivals of remembrance" and annually re-enacted the various scenes of her affliction. The language of bourgeois neurosis both appropriates and disavows, fears and longs for, a communal festive tradition no longer available.

From the seventeenth to the mid-nineteenth centuries we see literally thousands of acts of legislation introduced to eliminate carnival and popular festivity from European life. In different areas of Europe the pace varied, depending upon religious, class, and economic factors. But everywhere a fundamental ritual order of western culture came under attack – its feasting, violence, drinking, processions, fairs, wakes, rowdy spectacles, and outrageous clamor were subject to surveillance and repressive control. In 1855 the Great Donnybrook Fair of Dublin was abolished and the same year Saint Bartholomew's Fair in London was also abolished after the London City Missions Society petitioned for its suppression. In the decade following the Fairs Act of 1871, over 700 fairs, mops, and wakes were abolished in England. By the 1880s the Paris carnival was being rapidly transformed into a trade show-cum-civic/military parade (Bercé 1976) and as Wolfgang Hartmann has shown (Hartmann 1976), in Germany in the aftermath of the Franco-Prussian war, traditional processions and festivities were rapidly militarized and incorporated into the symbolism of the state. We now have available a range of good social histories of this process for most European societies and we can produce a sketchy but generally agreed map of the suppression of carnival. The impact and the extent of this long process is still underestimated. In Bavaria in 1770, for example, around 200 days a year were considered festive holidays. By 1830, sixty years later, under pressure from incipient capitalism in the region, this 200 days had been reduced to around 85 days a year (if Sundays are excluded this means a reduction of actual festivals from 150 to 35 a year). This dramatic collapse of the ritual calendar had implications not only for each social formation but for the basic structures of symbolic activity in Europe: carnival was now everywhere and nowhere.

Nearly every social historian treats the attack on carnival as a victory over popular culture by the absolutist state and then by the middle classes, a process which is viewed as the more or less complete destruction of popular festivity – the end of carnival. In this vision of the complete elimination of the whole ritual calendar there is the implicit assumption that, in so far as it was the culture of a rural population which was vanishing, the modernization of Europe led inevitably to the supersession of traditional festivity: it was simply one of many casualties in the move towards an urban industrial society. On the other hand, recent literary criticism has found elements of the carnivalesque everywhere it has looked in modern literature. Following the pioneering work of Bakhtin (1968), critics now discover the forms, symbols, rituals, and structures of carnival to be fundamental elements in the construction of modernism (White 1982). By and large, literary critics have not asked how or why this "carnivalesque" material should inform modern art because, busy with the task of textual analysis, they move too rapidly away from social practice to textual composition. Yet the social historians who have charted the demise of carnival as social practice, have not registered its *displacements* into bourgeois discourses, like art and psychoanalysis: adopting a naively empirical view, they have outlined a simple disappearance, the elimination of the "carnivalesque."

But carnival did not simply disappear. At least four different processes were involved in its ostensible break-up: fragmentation, marginalization, sublimation, and repression.

Carnival had always been a loose amalgam of procession, feasting, competition, and spectacle, combining different elements from a large repertoire in different localities. Even the great carnivals of Venice, Naples, Paris, and Nuremberg Schembart were fluid and changeable in their combinations of practices, though each had an enduring core of traditional ritual. During the long process of suppression there was a tendency for these mixtures to break down, with certain elements becoming separated from the others. Feasting became separated from performance, spectacle from procession. The grotesque body began to fragment. At the same time, it began to be marginalized both in terms of social class and geographical location. It is important to underscore the point that even as late as the nineteenth century, in some places, carnival remained a ritual that involved all classes and sections of a community, as it had once done for the whole of Europe. The disengaging of the middle class from it was a slow, uneven matter and part of that process was the "disowning" of carnival, and the gradual reconstruction of the concept of carnival as the culture of the Other. In fact, it was that very act of disavowal on the part of an emergent bourgeoisie that *made* carnival into the festival of the Other. It had not always been so, and in the eighteenth and nineteenth centuries we can see the social and geographical shifting of the festive culture to a liminoid position in middle-class life.

Addison (1953) charts many of these geographical marginalizations in the English context during the seventeenth and eighteenth centuries. Within a town, the fair, mop, wake, or carnival, which had once taken over the whole town and permitted no "outside" or "outsiders" to its rule, was confined and removed from the aristocratic and well-to-do neighborhoods. In the last years of the Bury St Edmund's Fair it was "banished from the aristocratic quarter of Angel Hill and confined to St. Mary's and St. James's squares" (Addison 1953: 163). In and around London,

> Both regular and irregular fairs were being steadily pushed from the centre outwards as London grew and the open spaces were built over. Greenwich and Stepney were the most popular at one time. Others – Croydon's for example – came to the fore later when railways extended the range of

> pleasure as well as the range of boredom, until towards the end of the 19th century London was encircled by these country fairs, some of which were, in fact, ancient charter fairs made popular by easier transport. . . . Most of them were regarded by the magistrates as nuisances, and sooner or later most of those without charters were suppressed. Yet such was the popularity of these country fairs round London that to suppress them in one place led inevitably to an outbreak elsewhere, and often where control was more difficult. As the legal adviser to the City Corporation had said in the 1730's "It is at all times difficult by law to put down the ancient customs and practices of the multitude." (Addison 1953: 100)

In England, the sites of carnival moved more and more to the coastal periphery, to the seaside. The development of Blackpool, Brighton, Clacton, Margate, and other seaside resorts reflects a process of liminality which, in different ways, was taking place across Europe as a whole. It may be argued that marginalization is a *result* of other, anterior processes of bourgeois repression, and this is indeed the case. However, it must be seen as a distinct and different historical tendency from the actual elimination of carnival.

There is a remarkable case of hysteria, documented by a Dr Schnyder from Berne (Schnyder 1912), which brings the process of marginalization in England and the symptoms of hysteria together, in a strange and telling way. Schnyder was influenced by Freud early in the latter's career, and wrote "Le Cas de Renata" as a contribution to the study of hysteria along broadly Freudian lines. Renata was 25 years old when she came to Schnyder suffering from hysterical symptoms, together with her brother, a priest, also diagnosed as an hysteric, and Schnyder treated them at the same time.

Renata (the "reborn," a name she gave herself after the successful termination of her condition) described how her hysteria began on a visit to Brighton and, though it is difficult for us to credit, she recounts how the onset of her vomiting, headaches, and anorexia was precipitated by seeing the holidaymakers on Brighton beach! Unable to cope with what Schnyder paraphrases as "le spectacle de la promiscuité des sexes offerte par une plage mondaine," Renata, raised as a strict Catholic, was forced to seek out a priest to help her cope with the deep threat which the "sexual promiscuity of the crowded beach" seemed to represent. The incident is doubly significant, in that it reveals both the special phobic power of carnivalesque or festive scenes for the hysteric, and, at the same time, nicely illustrates the way in which the marginalized forms of popular festivity suddenly re-emerge in the heart of bourgeois life as the very site of potential neurosis.

Bourgeois carnival is a contradiction in terms. Positive elements of the carnival tradition – such as the "regenerative power of laughter," the utopian and politically subversive aspects of folk culture and its social openness – become lost or perverted by the emergence of bourgeois individualism. Bakhtin is right to suggest that post-romantic culture is, to a considerable extent, subjectivized and interiorized, and on this account is related to private terror, isolation, and insanity rather than to robust, communal celebration. Bakhtin, however, does not give us a convincing explanation of this *sublimation* of carnival. Social historians, on the other hand, tend not to consider processes of sublimation at all: for them, carnival came to an end and that was all. Social historians tend not to believe in the return of the repressed. But a convincing map of the suppression of carnival involves tracing migrations, concealment, metamorphoses, fragmentations, internalization, and neurotic sublimations. The *disjecta membra* of the grotesque body of carnival found curious lodgement throughout the whole social order of bourgeois life.

By establishing the mediating ground between Bakhtin's work and that of social historians of carnival, a new domain opens out for investigation. Before this domain is thoroughly explored, one can only speculate that it will be found to contain more than the insignificant nomadic residue of a great ritual tradition. I have shown elsewhere how much modern "transgressive" literature, from Sade to Thomas Pynchon, transforms popular festive material into a "Carnival of the night" (White 1982). But on the basis of the striking similarity between the principal neuroses thematized in Freud, and the symbolic practices of carnival thematized by Bakhtin, I would hazard something further, along the lines explored in this paper. Bourgeois society from the seventeenth century through to the late nineteenth century systematically suppressed the material "body" of festivity once common to the entire society for a large part of every year. Carnival became confined to certain sites on the edge of town or on the litoral and was then largely dislodged altogether, no longer able to control the whole urban milieu during its reign. More important, perhaps, carnival was marginalized in time as well as spatially. The whole carnival calendar, which had structured the European year, reluctantly gave way to the working week, under the pressure of capitalist industrial work regimes. At the same time as this process severely regulated the body, it separated out the symbolic, discursive levels from the ritualistic and appropriated them in a free-floating and uneven way.

The result was a fantasy *bricolage*, unanchored in ritual and therefore set adrift from its firm location in the body, in calendar time, and ritual place. Dispersed across the territories of art, fantasy, and style, in flux, no longer bound by the strict timetable of the ritual year, these carnivalesque fragments have formed unstable discursive compounds, sometimes disruptive, sometimes therapeutic, within the very constitution of bourgeois subjectivity. Hysteria in the late nineteenth century was doubtless compounded in part by this material and that which had been excluded at the level of communal practice returned at the level of subjective articulation, as both phobia and fascination, in the individual patient. It is more than accidental that the major foci of carnival pleasure – food, dirt and mess, sex and extreme body movements – find their neurasthenic, unstable, and mimicked counterparts in the discourses of hysteria.

In her unfinished autobiography, Jean Rhys gives a perfect cameo of the phobic fascination afflicting bourgeois women in their exclusion from carnival as they were educated to be nice, clean, well-behaved little girls. Rhys writes (and the period is about 1900):

> The three days before Lent were carnival in Roseau. We couldn't dress up or join in but we could watch from the open window and not through the jalousies. There were gaily masked crowds with a band. Listening, I would think that I would give anything, anything to be able to dance like that, the life surged up to us sitting stiff and well-behaved, looking on. As usual my feelings were mixed, because I was afraid of the masks. (Rhys 1979: 42)

Rhys perfectly captures the three elements which are so tangibly connected in the case history of Frau Emmy von N., with which I began, when she too looked onto the alien masks with fear. Placed on the outside of the grotesque carnival body articulated as social pleasure and celebration, the female bourgeois subject introjects the spectacle as a *phobic representation*. Exclusion here entails the return of the excluded as desire and fear. In this case the link is made by the third term which is the regulation of the girl's body which Rhys describes as "stiff and well-behaved." Carnival was too disgusting for bourgeois life to endure. As well, it was a symbol of community and communality which the bourgeoisie had had to deny in order to emerge as a distinct and "proper" class. This class pursuit of purity, and the repudiation of revelling in mess, which was a central feature of carnival, was a fundamental aspect of bourgeois cultural identity.

Inversion and dirt

One of the essential ways of describing carnival focuses upon the *ritual inversions* which it habitually involves. The "reversible world" and the "world turned upside down" are phrases used to denote the way in which carnival inverts everyday customs, rules, and habits of the community. Hierarchies are inverted, kings become servants, boys become bishops, men dress as women and vice versa. The elements associated with the bottom part of the body (feet, knees, legs, buttocks, genitals, belly, anus) are given comic privilege over the spirit and the head. The "normal" rules of moral custom are overturned and license and indulgence become the rule: the body is granted a freedom in pleasure normally withheld from it, and obscenity of all kinds, from mild innuendo to orgiastic play and a robust reveling in mud, excrement, and all sorts of "filth," is sanctioned. Carnival gives symbolic and ritual play to impurity, to the mixed, heterodox, messy, excessive, and unfinished formalities of the body and social life. It displays, and takes as its sign, what Bakhtin call the "Grotesque body," a body which is distended, protuberant, fat, disproportionate, and open (marked by its large orifices, its gaping mouth and anus). The carnivalesque inversion then, nearly always involves a foregrounding and ritualized display of the fat, impure, open, fluid body, the body realized and celebrated as, precisely, "grotesque."

In every case of hysteria he looked at, Freud questioned the total repression of the grotesque body. We can precisely correlate the features of the carnival sign with the deepest phobias of the hysteric. They match, term for term. The hysteric, marked by "rigidity" and by a well-documented process of straightforward repression, suffers acutely from a terror of the grotesque body: dirt, fat, and the lower bodily stratum.

One of the grimmest sources of nausea for Frau Emmy is meatfat, the principal symbol of *mardi gras*, "Greasy Tuesday," of "carne levare," of the roasting meat which gave carnival its name. For centuries, fat dripping and suet were essential to carnival celebrations, clearly visible in the French opposition between *jours gras* and *jours maigre*, and usually personified as a fat man and a lean woman, the opposition between carnival and Lent. Hysterics privately re-enact the battle between carnival and Lent, a battle in which the anorexic figure for Lent – a figure represented as emaciated, a kill-joy, old, and female, a cold-blooded figure of humourless fasting and sexual abstinence – is invariably the victor. It is as if the hysteric has no mechanism for coping with the *mediation* of the grotesque body in everyday life, except by violent struggles of exclusion. In these violent struggles it is not surprising that fragmented elements from the carnival repertoire should be seized upon as weapons.

Strikingly, it is the woman's body which becomes the major battleground in the hysterical repression of the "grotesque form." The women hysterics, horrified by fat, associate their own developing or developed bodies with both dirt and fat. It is an affliction which may or may not coincide clinically with *anorexia nervosa* (Bruch 1965), but what is certain is that if the woman makes the unconscious identification of her own female form with the grotesque body under conditions which make this a negative, rather than a joyful and sensual identification, then she will enact a set of rituals closely following those formerly prescribed for Lent. Renata, in Schnyder's case study, like Anna O., changes her language in the face of unbearable ideas and has a catechism of terms which she recites in English (her native language is French): among these catechistic terms "Woman-Shame-Fat" is a particularly telling and poignant set, which she repeats to Schnyder. She remarks:

> J'avais honte d'être femme, en me plaçant au point de vue de la forme du corps. Je me demandais ce que l'homme pouvait bien penser de la femme. J'étais humiliée d'être femme et genée de sentir qu'on me regardait. Etre gras me paraissait particulièrement humiliant, parce que la forme du corps était plus accusée encore. J'aurais voulu qu'on ne distinguât rien du corps de la femme sous les vêtements. (Schnyder 1912: 218)

Renata is so appalled by the thought of her own body becoming fat that she gets up in the night and tightly straps herself into her corset to prevent her body from swelling. This nocturnal ritual reveals clearly how the hysteric enacts compulsive Lenten mortifications in the struggle against the female body signalling itself as incipiently carnivalized (Renata calls it "la guerre à tout sentiment, à toute sensation").

A common mechanism the hysteric employs in order to cope with the threat of the grotesque body is a top/bottom displacement, reversing, in an uncanny manner, the *ritual inversion* of the body found in carnival. In carnival the lower bodily stratum, exaggeratedly bulging and dirty, is made the source of comic play. In the hysteric, terrors associated with the lower bodily stratum are converted into symptoms of the top half of the body. Renata's vomiting is eventually traced to the feeling of pleasure which she has in her womb when she eats. She vomits, she says, to "repousser ces sensations." In a footnote to the *Dora* case, Freud discusses the displacement upwards of "catarrh" in a 12-year-old girl, transforming her vaginal discharge, about which she was deeply anxious, into a nervous cough (Freud 1905: 101 n.15). In the case of Dora herself, she transfers the "pressure of Herr K's embrace" from the lower to the upper part of her body. Furthermore, there is a curious passage in the *Dora* analysis where Freud points out a further, and more seriously damaging displacement in Dora's behavior. She, too, suffers from a leucorrheal discharge, and any intensification of this leads to an increase in her vomiting and refusal to eat. Freud's bland ascription of this to woman's "vanity" might well be the beginning of another, quite different reading of the case, one which would be more troubled than Freud was by the problem of purity and disgust with respect to the woman's body. Freud writes:

> The pride taken by women in the appearance of their genitals is quite a special feature of their vanity; and disorders of the genitals which they think calculated to inspire feelings of repugnance or even disgust have an incredible power of humiliating them, of lowering their self-esteem, and of making them irritable, sensitive and distrustful. An abnormal secretion of the mucous membrane of the vagina is looked upon as a source of disgust. (Freud 1905: 103)

We can put this another way: Dora's mother suffered from a fairly severe cleanliness fetish ("She . . . was occupied all day long in cleaning the house with its furniture and utensils and in keeping them clean – to such an extent as to make it almost impossible to use or enjoy them"), and this was a symptom of a widespread inability to cope with dirt and excrement in the middle-class household towards the end of the nineteenth century. "Disgust" is the operative term in hysteria, and contrary to Freud's express attempt to discredit this "path of association" as primary in the case studies, his reduction of the purity question to a single sexual source seems to me inadequate.

"Disgust" is the key category in the case histories and it is upon the category of "disgust" that much of the analysis of post-Renaissance carnival must focus. Carnivalesque practices were disgusting when measured by the emergent hegemonic standards of urban bourgeois life. Norbert Elias (1978) gives an overall, long-term context to this, whereby the festive body was cleaned up as a necessary symbolic correlate to the growth in civic power from the fifteenth century onwards. What Elias calls "the threshold of shame and embarrassment" shifted such that, over several centuries, aspects of the body, its secretions and fluids, which were once perfectly acceptable across the whole social formation, were first pushed down and then finally expelled from the culture. Greenblatt adds, in his brilliant article, "Filthy Rites":

> In this separation, the "lower bodily stratum" steadily loses any connection with anything other than the increasingly disreputable dreams of alchemists and cranks. Eventually, all of the body's products, except tears, become simply unmentionable in decent society (Greenblatt 1982: 10)

In the *Studies on Hysteria* Freud uses the resonant term "the agencies of disgust" to describe the forces arrayed against him in his struggle to cure hysterics. Those "agencies of disgust" are precisely the same agencies which, in their public form, mobilized civic and religious authorities against carnival. The connection is a charged and difficult one, but if it can be clarified then we have the outline of a thoroughly historicized, social understanding of mechanisms that Freud mistakenly thought to be more or less universal. Indeed, I would argue that the virtual elimination of traditional carnival as a real social practice in Europe, by the late nineteenth century, led to a *pathological* re-emergence of the carnivalesque in certain domains of bourgeois life. Hysteria, which as a social pathology "peaked" around the end of the nineteenth century, and declined rapidly thereafter (Veith 1965), marks a key moment in this process. In this period, bourgeois culture produced a compensatory range of peripheral "bohemias" which afforded "liminoid" positions of a kind approximating to those of traditional carnival, but from which "respectable women" were entirely excluded and which differed in certain respects from traditional carnival. The bourgeois bohemias, like surrealism and expressionism, took over in displaced form much of the inversion, grotesque body symbolism, festive ambivalence, and transgression that had once been the provenance of carnival.

I think we are now in a position to try and sketch a preliminary account of the kinship of certain hysterical manifestations and carnival. Privately and publicly, respectively, they correspond to the "staging" of the normally *repressed poles* of the fundamental binary structures underpinning a culture. Carnival was the repeated and periodic celebration of the grotesque body – fattening food, intoxicating drink, free sexuality, the lower bodily stratum, the heteroglot, and the topsy-turvy. All of these are at an opposite pole to the terms governing everyday work and are repressed for most of the time. Carnival allowed the society involved to mediate into periodic ritual the culturally structured "otherness" of its governing categories. We might call this process of periodic mediation *active reinforcement*: all the *opposite* terms to the structuring ones are given an actual and active staging during the festival. This contrasts strongly with hysteria, in which Freud detected the process of what he called "reactive reinforcement" as fundamental:

> Contrary thoughts are always closely connected with each other and are often paired off in such a way that the *one thought is exaggeratedly conscious while its counterpart is repressed and unconscious*. This relation between the two thoughts is an effect of the process of repression. For repression is often achieved by means of excessive reinforcement of the thought contrary to the one which is to be repressed. This process I call *reactive reinforcement*, and the thought which asserts itself exaggeratedly in consciousness and . . . cannot be removed I call a *reactive thought*. (Freud 1905: 72)

This remarkable passage is consonant with everything we know from anthropology and linguistics about the basic binaryism of symbolic functioning. Carnival belonged to a form of society, eliminated by bourgeois life, which had *communal* cultural mechanisms for the "staging" of the dyadic – opposite terms to those in operation in the sphere of daily production. In its elimination of these rituals, bourgeois life became more and more vulnerable to the unexpected and destabilizing emergence of the otherness of the grotesque body. As social mediation and its active reinforcement were gradually destroyed by a class reliant upon pure repression and therefore reactive reinforcement, the communal, social expression of the grotesque body (which is the body of all of us) was destroyed. Hysteria was a privately constructed staging, an attempted mediation, of a series of threatening othernesses which once had been ritually displayed and had now become "unconscious."

Theoretically, the use of the term "carnivalesque" for any bourgeois cultural form commits us to a sophisticated model of social history – a model much more Bakhtinian than that currently used by social historians. Yet the small army of literary critics now regularly describing modern cultural phenomena as "carnivalesque," seems to have "totalized" too quickly, to use Sartre's term. Abstracted from cultural history, the carnivalesque is located in bourgeois fictions and art forms without the central questioning being faced: what are the mechanisms and consequences of this bourgeois *sublimation* of a material, physical practice into purely *textual* semeosis? The assumption is that carnival may have disappeared from the streets but re-emerges in the privacy of the middle-class drawing room on a printed page and yet remains an equivalent *representation* to carnival. In fact, it seems that bourgeois art was hopelessly inadequate to the burden of acting as a semiotic substitute for the whole gamut of ritual practice, activity, discourse, structure, and process which carnival embodied. And the word "embodied" is symptomatic. Carnival was a symbolic, ritualistic practice of the *body* or it was nothing. In *Rabelais and his World* Bakhtin goes to great lengths to emphasize this:

> The carnivalesque crowd in the marketplace or in the streets is not merely a crowd. It is the people as a whole, but organized in their own way, the way of the people. It is outside of and contrary to all existing forms of the coercive socio-economic and political organization, which is suspended for the time of festivity.
>
> This festive organization of the crowd must be first of all concrete and sensual. Even the pressing throng, the physical contact of the bodies, acquires a certain meaning. The individual feels indissolubly part of the collectivity, a member of the people's mass body. In this whole the individual body ceases to a certain extent to be itself: it is possible, so to say, to exchange bodies, to be renewed (through the change of costume and mask). At the same time the people become aware of their sensual, material bodily unity and community. (Bakhtin 1968: 255)

What happens when a hegemonic group destroys the physical, ritual practices of a whole society and then endeavors to utilize the symbolism and purely discursive forms of those rituals for its own ends? To put it more pointedly, part of the very process of moving from a rich, physical culture of the social body through to textual representations has been necessarily a repression of the body through the "agencies of disgust." The modernist novel is a partial, selective sublimation of certain carnival practices and as such is deeply ambivalent, both comforting and disturbing a polite and decorous culture.

We can see this process of sublimation nicely expressed in relation to Freud's own life. In one of his early letters to his fiancé, Martha Bernays, he discusses a visit she had made to the Wandsbeck Fair. He agrees with her that the self-indulgence of the common people is "neither pleasant nor edifying." He goes on to add that their own pleasures, such as "an hour's chat nestling close to one's love" or "the reading of a book," have disabled them from participating in such common festivities.

For the rising middle classes it seemed as though all that messy, disruptive, violent nonsense of carnival was at last being done away with. In fact, even though the carnival was over, a strange carnivalesque diaspora had just begun.

References

Addison, William (1953) *English Fairs and Markets*. London: Batsford.

Bakhtin, Mikhail (1968) *Rabelais and his World*. Trans. H. Iswolsky. Cambridge, Mass.: MIT Press.

Barber, C.L. (1959) *Shakespeare's Festive Comedy*. Princeton: Princeton University Press.

Bercé, Yves-Marie (1976) *Fête et révolte*. Paris: Hachette.
Bruch, H. (1965) "The Psychiatric Differential Diagnosis of Anorexia Nervosa," in J.E. Meyer and H. Feldmann (eds) *Anorexia Nervosa*. Stuttgart: Georg Thieme Verlag: 70–86.
Elias, Norbert (1978) *The History of Manners: The Civilizing Process I*. Trans. E. Jephcott. New York: Pantheon.
Freud, Sigmund (1963a) [1905] *Dora: An Analysis of a Case of Hysteria*. Ed. Philip Rieff. New York: Macmillan.
——— (1963b) [1909] "General Remarks on Hysterical Attacks," in *Zeitschrift für Psychotherapie und medizinische Psychologie*, Bd. I. Trans. D. Bryan in *Dora: An Analysis of a Case of Hysteria*. New York: Macmillan: 153–7.
——— (1954) *Sigmund Freud's Letters to Wilhelm Fliess: Drafts and Notes*. Ed. Marie Bonaparte, Anna Freud, and Ernst Kris. New York: Basic.
Freud, Sigmund and Breuer, Joseph (1966) [1895] *Studies on Hysteria*. New York: Avon.
Greenblatt, Stephen (1982) "Filthy Rites." *Daedalus* 3(1): 1–16.
Hartmann, Wolfgang (1976) *Der Historische Festzug: Seine Entstehung und Entwicklung im 19. und 20. Jahrhundert*. Studien zur Kunst des Neunzehnten Jahrhunderts 35. Munich: Prestel Verlag.
Rhys, Jean (1979) *Smile Please – An Unfinished Autobiography*. Berkeley: Creative Arts Books.
Scheff, Thomas J. (1979) *Catharsis in Healing, Ritual and Drama*. Berkeley: University of California Press.
Schnyder, L. (1912) "Le Cas de Renata: Contribution à l'étude de l'hysterie." *Archives de psychologie* 12: 201–62.
Veith, Ilza (1965) *Hysteria, the History of a Disease*. Chicago: University of Chicago Press.
White, Allon (1982) "Pigs and Pierrots: The Politics of Transgression in Modern Fiction." *Raritan* 2(2): 51–70.
——— (1983) "'The Dismal Sacred Word': Academic Language and the Social Reproduction of Seriousness." *Literature, Teaching, Politics* 2: 4–15.

8

Violence and the liberal imagination

The representation of Hellenism in Matthew Arnold

Vassilis Lambropoulos

> Mr. Disraeli, treating Hellenic things with the scornful negligence natural to a Hebrew, said the other day in a well-known book, that our aristocratic class, the polite flower of the nation, were truly Hellenic in this respect among others, – that they cared nothing for letters and never read. (Arnold 1892: 1)

Culture and anarchy are two basic terms that regularly help define the spaces of public action and the range of creative opinions available to the contemporary (western) intellectual between the first and the last revolt in Paris (1789–1968). They are not necessarily oppositional, although they have been often perceived as such. But together they circumscribe a certain area of lofty concerns and self-reflective passions thriving in proud isolation from politics, religion, or science. Their combined problematic is by now familiar to us from explosive biographical cases (like those of Schiller, Pound, Byron, and Mayakovsky), artistic movements (like Surrealism), intellectual trends (like the Frankfurt School), legends (like Wagner in the Dresden barricades), or renunciations (like Rimbaud's). The problem is better known as Art and Revolution, but this is a narrow articulation whose clamorous, clever clarity elevates ideas to ideals. For a broader understanding, we may turn to philosophy, which has been obsessed with the problem: Locke, Rousseau, Fichte, Heidegger, Dewey, Derrida, Habermas – the list could continue and would probably not even exclude the analytic school. The unsettling problem of the culture–anarchy relationship has provided infinite inspiration to philosophers during the last three centuries. From Descartes to Gadamer, Lyotard, and Rorty, the persisting question has been: what are the tasks of culture in a world of anarchy, what is anarchy with or without culture? Some thinkers have taken sides openly, others have not. But the interaction of the two notions can be found to lie consistently at the foundations of their system, and support their pronouncements on physics and its prefixes.

Matthew Arnold even wrote a book on the subject in 1869, giving it as a title this pair of terms. Understandably, its success has been massive and its influence pervasive. No other work in the Anglo-American canon of criticism and aesthetics has enjoyed its popularity. Successive generations of scholars, theorists, and teachers wishfully succumb to its seductive advocacy of beauty, reason, and letters, or at least feel obliged to address themselves to the same issues. Every discussion of literature, art, or culture in general will take it under serious consideration and acknowledge the continuing relevance of its critical vocabulary. The intensity of the reverence is such that no other attempt has been made to deal with the same issue on a similar scale. To a disconcerting extent, Anglo-American literary and cultural criticism remains a series of laudatory or dissenting remarks dutifully submitted to the margins of that central book, deriving their agenda from its concise outline of culture's mission. Our political understanding itself is indebted to his social philosophy, and consequently our culture has been largely Arnold's.

The reasons for the book's appeal may be sought in two major strategies. The first is the close linking of the main terms. A disjunction is what would normally be expected: either culture or anarchy lie ahead for mankind. And indeed nowhere does Arnold indicate that the two may coexist. On the other hand, he refuses to oppose them directly, as if they are not antithetical. Thus we are invited to think of them together, to combine them in a mutual challenge, to let them engage each other: culture makes sense in light of the possibility of anarchy, and anarchy dictates only culture as a proper response. The two constitute the horizon of our future course and encode the alternatives of our civilization. If we oppose them, we have to exclude one; only if we see them in their interdependence may we realize that they also activate a certain productive dynamism in each other, in a mutual transformation that may express the best human potential. In this respect, Arnold knew much better than most of his commentators: he presented anarchy as an option that does not have to be annihilated but rather neutralized, assimilated, almost redeemed into culture.

Nevertheless, no invocation of redemption may ultimately avoid the dilemmatic logic of opposition. Arnold was able to preserve the tension between his two main terms without sacrificing one to another, without banning totally the joy of self-affirmation from the process of salvation. Yet, a vision of salvation had to be offered, and therefore a basic dichotomy to be established. And this brings us to the second major strategy, which was the spectacular allegorization of that dichotomy. By the mid-nineteenth century, the idea of a secular salvation had taken many forms, including those of progress, scientific knowledge, racial purification, class revolt, and Puritanism. But all these forms were too immediate, too tangible to inflame and keep alight popular imagination, not to mention the mystical needs of the intellectual sensibility. A grandiose scheme, larger than life and purer than history, was required for the depiction of the present and the revelation of the future. The sacred and the earthly had to be transposed to a higher, a transcendental level, and to be seen there in all their sublime majesty. Arnold suggested for this purpose an allegory amenable to multiple variations, which divided everything – man, the world, experience, society, history, knowledge – through two elemental forces: the Hebraic and the Hellenic.

As in many other cases, Arnold borrowed the terms and the basic idea from earlier sources, but through a dramatic rhetorical and ideological appropriation he turned them into integral and convincing parts of his own argument. Extensive discussions and a whole chapter, "Hebraism and Hellenism," are devoted to this distinction, which supports and illuminates the book. Its importance cannot be overestimated: it not only describes the "spirit of the age" and the tasks of the nation, but above all it provides the appropriate context for an adequate understanding of culture. Just to posit culture, Arnold needs this dichotomy more than anarchy, because it makes possible the space where the two terms of the title come into contact. Furthermore, the book shows how any discussion of literature, criticism, art, genius, nation, and the other cultural values of the modern state presupposes and requires an elaboration on the historical antithetical relation between the fundamental natural forces, the Hebraic and the Hellenic. More than anybody else before or after, Arnold brought this allegory of sin and salvation to the forefront of the debates of his age and reduced all of them to its binary form. By staging this conflict in the religious and national conscience he succeeded in presenting anarchy as just another cultural alternative.

I

Commentaries on *Culture and Anarchy* tend to concentrate almost exclusively on the first term of the title, although the book is demonstrably less a defense of culture than a homily against anarchy. What accounts for this unanimous response may be the obvious fact that culture has become a totally positive term, without a negative opposite, while anarchy forms a strong polarity with authority. Culture only denotes cultivation, growth, and development, while anarchy directly recalls and indirectly refers to authority; culture implies both process and product, but anarchy imbalance and dissolution; culture does, anarchy undoes. Arnold's commentators, willingly or unwillingly, respond to his call for action against the impending anarchy when they devote their best attention to the workings of culture: they work to deter the onslaught of anarchy, to prevent the disease from spreading. And yet the direction of the project is indicated in the semantic structure of its title: anarchy is not placed against its opposite, authority, but is paired with culture; the implication is that culture deters anarchy by supporting authority. The main argument, then, as expressed elliptically here, is that authority needs culture to survive. Under the emergency created by the imminent danger of anarchy, the dissolution of order and the collapse of authority, the defense by culture becomes the absolute priority. This basic concern inspires Arnold's project and should therefore be the starting point of any analysis. Although Arnold repeatedly defines and explains the notion of culture in great detail, he never offers a definition of anarchy. Not that the term in his time was clear or that its reality was familiar to all. His neglect seems systematic and aims at a certain vagueness that is allowed to engulf the word in ominous implications. By leaving its meaning unclear while alluding to its threat in various contexts, he places the anomalous experience of disorder at the threshold of civilization and in the advancing horizon of progress. Anarchy is the anguish of the wrong move: he can always conjure its spectre up when he needs to discipline our ambitions. Nothing is said explicitly about anarchy, so that readers may feel and fear it. From its different appearances (or rather apparitions), we may infer what it is, but we are never told. We are encouraged to hear but not look, to read the signs but not ask about the wall. Anarchy is the difference, the absence, the otherness of order: we recognize and know it through our fear of lack of security and authority.

Arnold's observations cover mainly phenomena of discord and dispersion in two spheres, the religious and the political. In the first, intimations of anarchy are discovered in criticisms of traditional dogma and practices, and in suspicions toward orthodoxy expressed in sectarian movements of English Puritans and Protestant (English and Scotch) Nonconformists, such as the organization of the Independents (Arnold 1971: 44) or recent attempts to disestablish the Irish Church (1971: 17). In these trends against the establishment of the Church Arnold detects a cult of dissent, an addiction to intolerance and hatred, and a pervasive divisiveness. His estimate is that the whole Nonconformist movement is taking the Church apart, shaking its foundations, undermining its authority, and dividing the nation.

Signs of a parallel situation are found in the political sphere too. Here the threat of anarchy is manifest in local claims, criticisms of the class system, "outbreaks of rowdyism" (1971: 62), "worship of freedom in and for itself" (ibid.), and generally in demands for individual liberty.

> More and more . . . are beginning to assert and put in practice an Englishman's right to do what he likes; his right to march where he likes, meet where he likes, enter where he likes, hoot as he likes, threaten as he likes, smash as he likes. All this, I say, tends to anarchy. (ibid.)

There is an obvious distinction here between the right and its practice: it is a good thing to have it but a negative one to see just about anybody exercise it. This worry is reinforced by the choice of verbs – meet, march, enter, hoot, threaten, smash: in their natural succession, they re-enact the procession of anarchy. The danger, claims Arnold in a most revealing example, is not the Irish Fenian, the alien and conquered papist who struggles for independence, because that resistance we can always crush, and for good reason, by brutal force; the real danger is the Hyde Park rioter, the Protestant Englishman of the working class who demands liberty, because "the question of questions for him, is a wages question'' (1971: 65). Arnold believes that he is not a revolutionary but a rough who ''has not yet quite found his groove and settled down to his work, and so he is just asserting his personal liberty a little, going where he likes, assembling where he likes, bawling as he likes, hustling as he likes'' (ibid.). The author's point is clear: the threat of anarchy comes not from the outsider, who can be repelled and punished, but from the insider who challenges the system. The choice and succession of verbs again warns against the self-affirmation that may lure every exercise of freedom into excess.

Still in the sphere of politics, but on a larger scale, Arnold expresses apprehension about the positivistic and Manichean ''ways of Jacobinism'' (1971: 52), with its absolute trust in reason, and about the supporters of an indiscriminate Liberalism, all of them advocating change without having a new realistic plan to suggest. Their attitudes are reflected in their blind trust in machinery, which is in fact ''the one concern of our actual politics'' (1971: 27). But technology will not prevent social unrest. Arnold concludes that faith in individual freedom and in industrial advancements, along with the rapid decline of the old religious devotion, are the most telling signs of an impending anarchy spreading in many areas of the personal and national life and threatening the established political and church institutions. Its overall picture, as painted in *Culture and Anarchy*, includes the critique of various forms of establishment, the degeneration of traditional authority, individualism in the name of freedom, and the development of a mechanical and materialist civilization. The social machine is out of order, as attested by the ''exclusive attention of ours to liberty, and of the relaxed habits of government which it has engendered'' (1971: 64). The unbalanced social condition is seen not only in the inability of the aristocrats to govern but also in the helplessness of the other classes to provide viable alternatives. The egocentric skepticism of modernity has corroded the pillars of supreme power.

> We have found that at the bottom of our present unsettled state, so full of the seeds of trouble, lies the notion of its being the prime right and happiness, for each of us, to affirm himself, and his ordinary self; to be doing, and to be doing freely and as he likes. We have found at the bottom of it the disbelief in right reason as a lawful authority. (1971: 121)

For Arnold, the problem lies with the modern individual of the middle and lower classes – with the sovereign subject, that is, the individual of the modern, post-revolutionary era. And the specific problem is his newly acquired and much celebrated freedom: how can it be channelled and controlled? How can it be rendered harmless for the establishment and its institutions? How can it be neutralized, cleansed of its centrifugal tendencies? Belief in authority has to be restored – but this authority has now to be democratic, and its believers should be the subjects comprising the modern political system. How then can the subject be subjected? How can freedom be administered? A new religion is obviously needed that will inspire trust, faith, and obedience – a religion of secular order that will help people resist the temptations of liberty and avoid the sin of anarchy. Arnold proposes a "religion of culture" (1971: 58), and as an ardent "believer" he seeks "to find some plain grounds on which a faith in culture . . . may rest securely" (1971: 32).

II

Culture is "a study of perfection" (1971: 34),

> a pursuit of total perfection by means of getting to know, on all the matters which most concern us, the best which has been thought and said in the world; and through this knowledge, turning a stream of fresh and free thought upon our stock notions and habits, which we now follow staunchly but mechanically. . . . And the culture we recommend is, above all, an inward operation. (1971: 5–6)

Arnold promulgates "the idea of perfection as an *inward* condition of the mind and spirit," "as a general expansion of the human family," and "as a harmonious expansion of human nature" (1971: 38). His dogma replaces salvation with perfection, faith with knowledge, the transcendental with the inside, elevation with growth, purification with harmony, atonement with cultivation. He argues that "of perfection as pursued by culture, beauty and intelligence, or, in other words, sweetness and light, are the main characters" (1971: 58). His idea of culture is a private pursuit of beauty and intelligence by means of knowing the canon of our tradition – an inward operation of growth that leads to harmony and fulfillment. He offers this vision of personal cultivation and development to those who claim free exercise of their rights as an alternative goal in life and as a program for national survival – to the philistines of the middle class now in power and to the populace demanding equality.

Arnold's idea of culture as a religion for the modern undereducated or underprivileged masses is based directly on the principles of aesthetics. He presents it as "the disinterested endeavour after man's perfection" (1971: 22) that is free of materialist concerns and instead focuses on the mental faculties. "Culture, disinterestedly seeking in its aim at perfection to see things as they really are" (1971: 24), helps the intellect grow freely in harmony and knowledge. He outlines the "free spontaneous play of consciousness" (1971: 167) whereby the subject pursues pure intellectual fulfillment through the grasp of knowledge and the contemplation of beauty. Repeated references to beauty, freeplay, disinterestedness, harmony, perfection, inwardness, purity, and autonomy show that Arnold's ideal of "culture and totality" (1971: 15) is an aesthetic one, and his praise of culture a strong defense of high art.

He establishes the parallel explicitly: ''In thus making sweetness and light to be characters of perfection, culture is of like spirit with poetry, follows one law with poetry'' (1971: 42). Culture propagates life as art, art as poetry, and poetry as writing. Culture, the disinterested contemplation of beauty, is the secular cult of disinterestedness, which ultimately promulgates leisure as writing, but also writing as leisure. Arnold is preaching to the bourgeoisie and addressing himself to their needs and worries – the demands for more profits and rights, the spiritual quest, education, training for government, exercise of power. All this is skillfully sublimated in the ideas of inwardness, disinterestedness, play, and perfection. Modern individualism can be both celebrated and controlled by being encouraged to celebrate its self – in purity, autonomy, harmony.

Any discussion of *Culture and Anarchy* should not lose the perspective provided by the subtitle of the book: ''An Essay in Political and Social Criticism.'' Arnold's defense of culture is only part of a critique directed against the anarchy looming over society and its institutions. Therefore invocations of beauty and appeals to intelligence should be understood as strategic employments of notions that are intended to counter disorder. The pursuit of culture may be disinterested, but its uses serve (and are expected to continue to serve) very specific social and political interests. Arnold is careful to stress that culture, although a freeplay for the individual, is his duty towards society. It should therefore be ''considered not merely as the endeavour to *see* and *learn* this [things as they really are], but as the endeavour, also, to make it *prevail*'' (1971: 36). The nation needs the enlightenment that only it can provide with its beneficial social/moral effects. ''Now, then, is the moment for culture to be of service, culture which believes in making reason and the will of God prevail'' (ibid.). Arnold is clearly talking not only about edification but also about real power.

Culture, in order to acquire power and prevail, should be broadly distributed. The old means of administration by coercion were in the hands of the few; culture, however, should be the privilege of the masses, and therefore made widely available in order to effect administration by subjection. All subjects must go through a process of enculturation. Sweetness and light are not enough, and ''we must have a broad basis, must have sweetness and light for as many as possible'' (1971: 55). Culture, as a bourgeois ideal, is egalitarian and liberal: it advocates equality in perfectibility in a democracy of independent subjects.

> It seeks to do away with classes; to make the best that has been thought and known in the world current everywhere; to make all men live in an atmosphere of sweetness and light, where they may use ideas, as it uses them itself, freely, – nourished, and not bound by them. This is the *social idea*; and the men of culture are the true apostles of equality. (1971: 56)

In both its attempt to permeate society and prevail, and in its egalitarian attitude, culture again strongly resembles religion. "Not a having and a resting, but a growing and a becoming, is the character of perfection as culture conceives it; and here, too, it coincides with religion" (1971: 37). In regard to our duty to spread the message, "culture lays on us the same obligation as religion" (ibid.). The only substantial difference is that the former replaces faith with knowledge. "What distinguishes culture is, that it is possessed by the scientific passion as well as by the passion of doing good" (1971: 35). This is, then, the secular religion of the bourgeois era: democratic, egalitarian, disinterested, open-minded, expansive, it makes pure beauty and correct knowledge a common good and cause, protecting from public anarchy through private perfection.

Thus in Arnold's scheme culture is essentially a means for preserving, improving, and enhancing authority. For this reason, its subject is the canon itself, the respected tradition of acknowledged masterpieces.

> The great works by which, not only in literature, art, and science generally, but in religion itself, the human spirit has manifested its approaches to totality and to a full, harmonious perfection, and by which it stimulates and helps forward the world's general perfection, come, not from Nonconformists, but from men who either belong to Establishments or have been trained in them. (1971: 9)

As we know, "establishments tend to give us a sense of a historical life of the human spirit, outside and beyond our own fancies and feelings" (1971: 16). Culture is the music produced by the operations of those establishments, the melody of their effective functions. In turn, its social task is to serve, to strengthen them. By its origin, culture is canonical, inextricably bound to authority. The establishment is the source and its master, authority its proper realm of development. When culture prevails, authority prevails as well; when people become encultured, they are simultaneously subjected to rational control. Culture gives authority the ultimate justification – inherent value; in addition, by training subjects, it makes coercive subjection redundant.

III

Culture, however, works both ways: it not only trains people for authority but also contributes to the reformation and modernization of authority. In regard to the latter, Arnold argues, culture is now suggesting a new model and center, the state. In a period of expansion and decentralization, innovations and revisions may happen too fast for the establishment to adjust. "Everywhere we see the beginnings of confusion, and we want a clue to some sound order and authority" (1971: 120). There is an urgent need for new organization and structure, for an orderly transformation of the old system that will contribute to what was called a "revolution by due course of law" (1971: 173). Only culture, with its sweet light, may point the way to which "the assertion of our freedom is to be subordinated" (1971: 63). That authority, whose rules will be right reason, the will of God, and one's best self, is the modern state. Arnold closes his book by expressing the conviction that "our main business at the present moment is not so much to work away at certain crude reforms of which we have already the scheme in our own mind, as to create, through the help of that culture which at the very outset we began by praising and recommending, a frame of mind out of which the schemes of really fruitful reforms may with time grow" (1971: 167–8). Priority is given not to the system but to mind: reform must first be executed there. During the continuing decline of traditional authority, the state, through the appropriate cultural training, must distract the growing skepticism of the masses for the existing forms of establishment, and instead educate them, by disseminating the artistic and literary canon, in the pleasures and duties of subjectivity. Division, Nonconformism, disbelief, criticism, and protest, which are threatening the foundations of the class and Church systems, ought to be diverted by and into culture, while authority will undergo a transformation into a benevolent state. "Well, then, what if we tried to rise above the idea of class to the idea of the whole community, *the State*, and to find our centre of light and authority there?" (1971: 77). People are threatened by selfishness, the greed of their common, everyday selves, and a supreme authority is needed to save them from themselves. "We want an authority, and we find nothing but jealous classes, checks, and a deadblock; culture suggests the idea of *the State*. We find no basis for a firm State-power in our ordinary selves; culture suggests one to us in our *best self*" (1971: 78).

According to Arnold, then, the three steps that man has to take, if he is not to fall into chaos, are culture, self, and state. In his terminology, beauty and knowledge will cultivate right reason and apply it to the establishment of an enlightened, democratic authority. That authority is the "*State*, the power most representing the right reason of the nation, and most worthy, therefore, of ruling, – of exercising, when circumstances require it, authority over us all" (1971: 67). As this model of order shows, the state represents the rules, or, rather, rules by representing: it re-presents the people *as* a nation, and their rights *as* government. The government of the nation is the representation of the educated citizens, of the acculturated subjects, of that state. Authority is now turning into an exercise more of control than of force. Individual demands and practices of rights will be rendered meaningless by this highest manifestation of communal solidarity and desire which is already expressed in the "idea of a *State*, of the nation in its collective and corporate character controlling, as government, the free swing of this or that one of its members in the name of the higher reason of all of them, his own as well as that of others" (1971: 65). Thus the state, like the best, the individual self, will be built on the model of the autonomous, self-sufficient, radiant artwork and based on aesthetic principles suggested by culture. It is time for people to create, and assimilate themselves into, the state of culture.

As we saw earlier, while the inward operations of culture are not of social relevance, their repercussions have major political importance. Activities within the realm of culture may be private and disinterested but they eventually affect the establishment. The relationship between culture and authority, however, is mutually constitutive and beneficial in that at the same time the latter sanctions and consecrates the former. As we can see clearly in the case of the state, culture and authority produce and support each other. In the state projected by Arnold, in this promised land of the bourgeoisie, culture, on the one hand, authorizes the validity of the communal consensus, while the state, on the other, recognizes it as its official faith and makes it the national religion of the secular kingdom. In this context, the role of culture is therapeutic: it cures people of their worst selves and authority of its unnecessary excesses, restoring health, reason, unity, sweetness, and light in the shaken establishments. But for a better understanding of its foundations, we ought to inquire deeper and "find, beneath our actual habits and practice, the very ground and cause out of which they spring" (1971: 106). And the ground and the cause of our being and plight is the double, schismatic, contradictory identity of our civilization – Hebraism and Hellenism.

IV

For almost three centuries, since it first posited itself as an essence, a problem, and a quest, western thought has conceived of the world in terms of an all-encompassing polarity: the Hebraic vs. the Hellenic. In literature, art, criticism, scholarship, epistemology, and metaphysics, in different forms and manifestations, whenever thought has portrayed itself (as conscience, subconscious, inspiration, knowledge, talent, language, Being, writing, subjectivity, fragmentariness, or gender) and has inquired after non-thought (the negative, God, difference, otherness, alienation, silence, lack, absence), it has always operated on the basis of this antithesis, where the Hebraic is the positive term – the depth, the horizon, and the meaning – while the Hellenic is the opposite – the surface, the moment, and the message. Thought thematizes and articulates itself as the Hebraic, the dark silence of the ontic; and it questions its materiality, its fleeting presence in this Greek world of blinding light and deceptive form. Ever since the western mind asked the question of identity, from Spinoza to Derrida, it has been searching for the different – because the question of identity is the search for the secular transcendence of difference. And the different is always Hebraic – muted, strange, exilic, always already chosen and punished: chosen to sin and punished to be chosen; while identity is the source and the cause of guilt – the material, the profane, the present, the temporary, the exchange value of the sign, the never before or again. The Hebraic stands for the transcendent: thought about itself and man against his (and lately her) self; while the Greek stands for the worldly, the earthly, the limited, the finitude of use. Ultimately, the distinction is between the Old Athens and the New Jerusalem, the Acropolis and the Temple, the philosopher and the prophet, beauty and faith, perfection and salvation. But as in every dialectical scheme, the two need, define, verify, support, reinforce each other: for thought to seek its other it must posit a self, to find salvation it must indulge in sin, to find meaning it must create form.

If Arnold's treatment of the polarity has been the most popular, this may be attributed partly to two reasons: his strong identification of the Hebraic with authority, and his vision of a possible reconciliation of the opposite forces. According to his program of national rejuvenation, the role of culture will be to renew and strengthen authority by preparing and cultivating individuals for a new national consensus to be expressed and monumentalized in the state. The old order will not be destroyed (let alone allowed to collapse) but rather supplemented by a new force, so that eventually a reconciliation of classes, religions, denominations, individual interests, and political goals will be implemented. In this scheme, the two forces of order and renovation are represented (and allegorized) respectively by Hebraism and Hellenism; and the remedy for the national malady is rational, orderly, educated, and informed Hellenization of the Hebraic order that has grown old, fanatic, and exclusive.

Arnold endorses the traditional absolute distinction between them: "Hebraism and Hellenism – between these two points of influence moves our world" (1971: 107). But he is careful not to accept their opposition as necessary: "And these two forces we may regard as in some sense rivals, – rivals not by the necessity of their own nature, but as exhibited in man and his history, – and rivals dividing the empire of the world between them" (ibid.). On the other hand, this situation should not continue, since the two trends have a major element in common: "The final aim of both Hebraism and Hellenism, as of all great spiritual disciplines, is no doubt the same: man's perfection or salvation" (1971: 108). In this sentence, especially the last parallelism, Arnold may have captured the essence of the whole dichotomy in its fundamental isomorphism: profane and divine, secular and holy, art and religion, beauty and faith – the equation of the two shows their ideological roots in the search for transcendence in both worlds, this and the other (– any other). Salvation and perfection, the Hebraic and the Hellenic, are the two sides of the dialectic coin. "At the bottom of both the Greek and the Hebrew notion [of felicity] is the desire, native in man, for reason and the will of God, the feeling after the universal order, – in a word, the love of God" (1971: 109). The forces may be radically different but they are isomorphically parallel, and their efforts converge in the search for God's love as the law of universal order. If we can work toward making them converge on this earth and in this life, Arnold suggests, we shall establish, in the absolute, all-embracing institution of the state, the law of worldly order. And the model of perfection provided by art, as encoded in culture, will help us achieve secular salvation in the balanced, total order of communal will, reason, and desire – the state.

Distinctions between the trends are not and cannot, of course, be eliminated. "They are, truly, borne towards the same goal; but the currents which bear them are infinitely different" (1971: 110). Their approaches differ greatly. This, however, makes them complementary rather than antithetical:

> their single history is not the whole history of man; whereas their admirers are always apt to make it stand for the whole history. Hebraism and Hellenism are, neither of them, the *law* of human development, as their admirers are prone to make them; they are, each of them, *contributions* to human development, – august contributions, invaluable contributions. (1971: 115)

But Arnold urges his audience to work toward reconciliation and combination, a final, stable synthesis. He envisions a world where "man's two great natural forces, Hebraism and Hellenism, will no longer be dissociated and rival, but will be a joint force of right thinking and strong doing to carry him on towards perfection" (1971: 173). This is his dream of totality and integration of the two elemental forces, after their conflicts in history have been overcome: reconciliation and transcendence, because man needs them both. The spread of culture will help this happen, and it will take place in/as the institution of state. But before we elaborate on that, we ought to explain his conception of the two powers.

The Hebraic belongs to, and expresses, the realm of the moral: it represents doing, acting, and believing. It provides principles of behavior and rules of conduct, and commands obedience to them. The Hellenic belongs to, and expresses, the realm of the intellectual: it represents thinking, knowing, and exploring. It provides light and beauty, and inspires spontaneity. The first emanates from the social and the public, while the other from the personal and the private. Therefore the two define different spheres of experience, both basic and important. To generalize further, the Hebraic deals with issues of the soul, and the Hellenic with issues of the mind. And they find their most paradigmatic expression and systematization in religion and art respectively. This is in fact what each civilization contributed to humanity: our religion and faith are Jewish, our art and beauty Greek; we owe salvation to the former, perfection to the latter. Undoubtedly, religion is the more important experience; yet it is often stark, unsettling, demanding – it may even lead, in its fanatic expressions, to hostility to man, because self-conquest requires severe moral strictness. That is why it needs the clarity, simplicity, and freedom of art provided by Hellenism in the comprehensive expressions of culture. Virtue needs to be balanced with disinterestedness, obedience must be sweetened with play. Extremes should meet, negotiate, and merge harmoniously according to the laws of their nature.

Hebraism and Hellenism, then, must be properly combined and balanced. That is not the case today, though, and the resulting disequilibrium has allowed anarchy to develop into a real threat. Authority, Arnold suggests, must again be balanced and its exclusiveness tempered. Historical developments, he proposes, have led to too much Hebraization. In his general view of history,

> by alternations of Hebraism and Hellenism, of a man's intellectual and moral impulses, of the effort to see things as they really are, and the effort to win peace by self-conquest, the human spirit proceeds; and each of these two forces has its appointed hour of culmination and seasons of rule (1971: 116)

Thus in the sixteenth century Hellenism re-entered the world with the Renaissance after centuries of disappearance following the Hebraic triumph of Christianity. Then came a Hebraizing revival, the Reformation and the return to the Bible, which culminated in Puritanism. Arnold finds that this stage persists. "Obviously, with us, it is usually Hellenism which is thus reduced to minister to the triumph of Hebraism" (1971: 108). He agrees with Ernest Renan (1868) that this phenomenon has reached its extremest form with American Puritanism. "From Maine to Florida, and back again, all America Hebraises" (1971: 15). Therefore the current neglect of Hellenism may be attributed to "the long exclusive predominance of Hebraism" (1971: 130), which accounts for the loss of balance in the exercises of authority. Now it is deemed necessary to cure Puritanism of its excesses and redress the balance. Moral and religious feelings will not survive the onslaught of skepticism unless buttressed with the intellectual pursuits of culture. Faith must be fortified with knowledge and cultivated as an art.

Arnold's vision of an enlightened religiosity directly addresses this pressing issue, the rejuvenation of Hebraic authority with Hellenic culture. The roots of Christianity, he argues, are both Jewish and Greek. Therefore, a Christian nation should be organized as a system that draws from these two sources. In the present stage, Hellenism should receive more attention and encouragement. "Now, and for us, it is time to Hellenise, and to praise knowing; for we have Hebraised too much, and have over-valued doing" (1971: 27). Arnold is not, of course, advocating complete Hellenization, but only a minor (yet indispensable) adjustment of the existing system.

> And when, by our Hebraising, we neither do what the better mind of statesmen prompted them to do, nor win the affections of the people we want to conciliate, nor yet reduce the opposition of our adversaries but

> rather heighten it, surely it may not be unreasonable to Hellenise a little, to let our thought and consciousness play freely about our proposed operation and its motives, dissolve these motives if they are unsound, – which certainly they have some appearance, at any rate, of being, – and create in their stead, if they are, a set of sounder and more persuasive motives conducing to a more solid operation (1971: 145–6)

This is the gospel of Arnold's liberalism: if authority is losing ground, if division and animosity divide people and nations, then some freeplay, some informed rethinking, and some flexibility may help the reinforcement of law and order. Let us be more open-minded about administration and government, Arnold advises: when he defends the idea "to Hellenise, as we say, a little" (1971: 167), he simply means "the habit of fixing our mind upon the intelligible law of things" (1971: 166), of supporting religion with evidence and argument now that faith is far from able to guarantee the old habits of obedience. He is essentially defending the right of the new individual, the subject, to entertain his own thoughts at leisure and free of interest, when properly trained and informed: "Plain thoughts of this kind are surely the spontaneous product of our consciousness, when it is allowed to play freely and disinterestedly upon the actual facts of our social condition, and upon our stock notions and stock habits in respect to it" (ibid.). His principal concern is how the aesthetic attitude and behavior will be integrated in bourgeois life, at the same time fulfilling and justifying it. For this purpose, and for the ultimate task of the preservation of authority in mind, it would suffice to effect a "fruitful Hellenising within the limits of Hebraism itself" (1971: 154). Thus Hellenism will supplement, temper, and balance the excesses of Hebraism, and will preserve its essence intact.

Arnold makes an intense plea for "mutual understanding and balance" (1971: 122) between the two supreme forces. Still, throughout his discussion and his defense of Hellenism, he leaves no doubt about its subservient role and the superiority of Hebraism. Notice how he does not lose sight of his priorities when he admonishes that "we are to join Hebraism, strictures of the moral conscience, and manful walking by the best light we have, together with Hellenism, inculcate both, and rehearse the praises of both. Or, rather, we may praise both in conjunction, but we must be careful to praise Hebraism most" (1971: 123). This is an explicit warning: the Hellenic spirit in itself is unimportant, knowledge, culture, and the disinterested pleasures of subjectivity are meaningless, unless they serve the Hebraic establishment. Hellenism, through culture, will help "the diseased spirit of our time" (1971: 137) survive, and explore new ground for, and forms of, authority. "Hellenism may thus actually serve to further the designs of Hebraism" (1971: 133). Culture is necessary only to the extent that it protects and advances the causes of authority, whose foundations lie in the Jewish faith.

V

In writing *Culture and Anarchy*, Arnold is presenting a concrete political project: he is outlining a method for salvaging the existing modes of authority from the danger of anarchy by renewing order. He realizes that the old order has lost its credibility; he therefore looks for means of overcoming the rigidity and suffocation, and inspiring again confidence and obedience. Faced with voices publicly demanding change, exercising criticism, and claiming rights, he counter-argues by shifting the terms of the debate. For him the problem is not one of structure but of arrangement; not of default but of balance; not of power but of authority; not of politics but of science; not of religion but of culture; not of failure but of efficiency. His main aim is to fight the modern, the revisionary, the radical; he fights the combative with the dialectic. The main strategy consists in presenting the new, the unknown, the critical, as non-new, in fact as only the other, neglected half of the old. Religion and culture, he argues, together form the basis of our civilization: they represent faith and intellect respectively, and they need and entail each other. Naturally, the one is more basic and important than the other; but both are by definition implicated in a good and balanced society. The authority of religion needs the enlightenment of culture. When either predominates, the seeds of anarchy are sown. For Arnold, then, the contemporary problem is not the establishment but its excesses and abuses. He believes that cultivated, right reason may prevent them. For this reason, he seeks a strategic adjustment of power, one that will allow more individuals – namely, the participants of culture – to take part in its administration by volunteering their grateful submission. The liberalization of power will protect it from libertarian demands, and the modernization of authority will guard it against the modern. More power will not help establishments achieve subjection; instead, culture will cultivate subjects – disciplinary topics made for knowledge will educate disciplined individuals for authority.

The distinction between authority and culture is allegorized by Arnold in the Hebraic–Hellenic dichotomy. Hebraism represents authority at its best, most meaningful, and most enduring – religion. It gives it an apocryphal background, a timeless relevance, a transhistorical validity, and a prophetic power. Hellenism, on the other hand, represents enlightenment at its most comprehensive and consummate – culture. It gives authority a sense of tradition, a continuous past, a glorious history. The problem facing Arnold is how to protect religion from political criticism, and politics from individual intervention. To the threat of this problem he gives the name of anarchy, or lack of authority. His solution is to endow authority with the prestige of culture, religion with the wisdom of right knowledge, and to transform the esoteric, exclusive system of the establishment into the public, panoptically present institution of the state. The concrete suggestion, in terms of cultural enrichment, is to look at the "best art and poetry of the Greeks, in which religion and poetry are one, in which the idea of beauty and of a human nature perfect on all sides adds to itself a religious and devout energy" (1971: 43), and to imitate that model, to try to reach again that harmonious fusion. He observes that, in the modern world, literature, religion, and politics suffer from a severe "absence of any authoritative centre" (1971: 91). Religion and art/poetry have been of course separated, and the former is losing its justification while the latter is gaining in both independence and respect. But the center does not hold, and perfection is growing into an isolated aesthetic pursuit. "We have most of us little idea of a high standard to choose our guides by, of a great and profound spirit which is an authority while inferior spirits are none" (1971: 92). We should therefore look to art for a model and aspire to a broad culture which will provide "a certain ideal centre of correct information, taste, and intelligence" (1971: 91).

Culture is called upon to support, enlighten, and justify authority – not in its old form, though, the religion of authority, but in a new one: the art of authority. Until now, religion and its strict principles of conduct was the model of authority, as encapsulated in the Hebraic; but now, for the survival of the power of the establishment, a new model is needed, that of culture-as-art exemplified by the Hellenic. The Hebraic authority of religion, based on rules of obedience and belief, ought to be supplemented by the Hellenic authority of culture, which invites obedience as disinterested knowledge. Although authority and culture, religion and art, have so far been rivals and antagonistic forces, it is now time to be reconciled, because they can beneficially complement each other. Culture informs, enlightens, and serves authority, but without its grounding it is meaningless, if not impossible; religion, on the other hand, needs the sweetness and light of art. Culture is an auxiliary yet indispensable instrument of power, and this holds true for the Hellenic in its relationship with the Hebraic. Arnold argues that authority needs culture, religion art, virtue beauty, consciousness thought, the tyrant the poet, and power truth. In his vision of the new, all-embracing, benevolent establishment of democratic administration, the state, art will be the new faith and culture its religion: culture, as the religion of the state, will be the first secular dogma for all people, in which everybody can play freely and spontaneously. Culture will be the religion, the dogma, and the morality of the bourgeoisie.

Arnold's conception of culture-as-art, as the religion of the modern era, the cult of the dominant middle class, dictates two major principles and requires two corresponding socio-historical developments. The first is the state as the supreme form of political authority. Building it is an urgent priority because

> a State in which law is authoritative and sovereign, a firm and settled course of public order, is requisite if man is to bring to maturity anything precious and lasting now, or to found anything precious and lasting for the future. Thus, in our eyes, the very framework and exterior order of the

> State, whoever may administer the State, is sacred; and culture is the most resolute enemy of anarchy, because of the great hopes and designs for the State which culture teaches us to nourish. (1971: 170)

The state, then, is Hebraic authority fortified with Hellenic culture. It is characterized by sovereignty, order, direction, maturity, permanence, but above all a sacred form. In its case, the content or the agent of power are not as important as its surface and appearance as authorized by culture, because the state is pure form and total signification: authority as art. The state of culture is a product of the art, not the religion, of authority: it commands by soliciting perfection. The order of the state is sacred because it is the order of culture, of art, of form, of harmony, of the independent signifier, of the naturalized language, of pure writing, of the disinterested play with difference.

The second major principle and requirement dictated by Arnold's conception of culture is perfect individuality, the best self the bourgeois may submit for the approval of his rights. As he states in detail,

> what we seek is the Philistine's [the middle class's] perfection, the development of his best self, not mere liberty for his ordinary self. And we no more allow absolute validity to his stock maxim, *Liberty is the law of human life*, than we allow it to the opposite maxim, which is just as true, *Renouncement is the law of human life*. For we know that the only perfect freedom is, as our religion says, a service; not a service to any stock maxim, but an elevation for our best self, and a harmonising in subordination to this, and to the idea of a perfected humanity, all the multitudinous, turbulent, and blind impulses of our ordinary selves. (1971: 153)

The crucial terms in this passage form a recognizable set (renouncement, subordination, service) and are together opposed to liberty. The task of the bourgeois is to improve and perfect his self, and that, as we saw before, he is encouraged to perform in the most disinterested way. But the outcome must serve the interests of a perfect, harmonious humanity. As with art, interest is no longer part of the process but a component of the result. Thus perfection is an aesthetic, private procedure that produces a political, public result. The Philistine bourgeois must mould himself into an artwork – perfect, total, autonomous, independent, fulfilled, asocial, transhistorical. But this perfect self must submit his harmonious independence to a social service. Perfection, as an aesthetic goal, justifies itself, but liberty, as a political one, does not. As Hebrew religion has taught, freedom is submission; and as Hellenic art has shown, freedom is perfection. The modern individual, then, as an independent subject, should be willing to perfect and submit himself for voluntary service. His freedom requires (and is based on) renouncement – of the ordinary, the everyday, the commonplace, the non-artistic in general.

The individual self, the absolute expression of the middle class, is the common man purged of all commonness, absolved of all interests, who elevates himself to the level of art and serves the order of the state. His secular faith in culture generates his trust in worldly authority. The more independent and perfect he is, the better servant of the state he may become. His harmony with his self facilitates the harmonization of his pursuits with those of the state in that, through perfection, he subjects himself willingly. His highest ideals are a life of art and a state of culture. In Arnold's vision, the individual as artwork is the masterpiece of culture, and the state as site its museum, which succeeds the institution of the Church. From a matter of taste, the canon becomes a guide to good conduct. In this universe of decorated exteriors, natural signs, and self-conscious enjoyments, control reaches an incomparable purity as it turns into the sheer formalism of administration. The self at its best, a law of art unto himself, has been prepared to serve the rule of the state. Independence will cost him individuality, his rights will cost him his protests, and his culture his liberty. Art will administer his pleasure, culture his desire, science his thought, and the state his salvation. Selves are for service, subjects for subjection. Those who cannot be served by religion alone any more will be given another chance, this time on the personal level, to atone by/in/as art: they may renounce life and redeem the ordinary in beauty.

The Hebraism–Hellenism dichotomy that Arnold employs helps him dramatize the conflict between authority and dissent, and show a path toward reconciliation. The dramatization is effected through an opposition and allegorization of the two forces which are hypostasized as two natural powers, unraveling in (and constituting through their competition) western history. The Hebraic represents the dark kingdom of religion, while the Hellenic the radiant presence of art. In Arnold's scheme, the former stands for the moral austerity of order and authority, and the latter for the disinterested knowledge of harmony and culture. Culture is of course the first word in the title of the book. But as we noted earlier, the other basic notion, authority, appears in the title only through its semantic opposite, anarchy, and thus it is obscured – "made dark", i.e., Hebraic. The effect of this choice of words may be explained in two ways. One is to say that Arnold introduces a new polarity, implying that anarchy actually opposes culture and that, conversely, only culture can save us from anarchy. This view is partly corroborated by the original title of the book's first chapter, "Sweetness and Light," which, when the text was presented as a lecture, was "Culture and its Enemies." There is another possible reading, though, which may not be incompatible with this one. Let us again recall that the Hellenic represents mind, thought, exploration, spontaneity, and independence, and not soul, faith, morality, obedience, proper conduct, and strictness. As these polar distinctions imply, the Hellenic is not only what generates culture, but also what makes anarchy possible. If the Hebraic is order and the Hellenic play;

if order requires faith and play facilitates inquiry; if the faith in order produces obedience and the play of inquiry advances knowledge (whose consequence is sin) – then the power of the Hellenic is truly ambiguous and ambivalent: it may lead to either renewed or to overturned order, it may strengthen authority or engineer anarchy, it may release either the beneficial or the eruptive power of culture. Anarchy, then, is part of the Hellenic potential – it is the uncontrolled, untamed, free, skeptical, and irreverent Hellenic.

Arnold seems to realize that exclusive, static, coercive power provokes extreme critiques of authority, and invites disobedience with its traditionalism. His method of defense includes three tactics: first, to allegorize the two forces by branding them with the names and emblems of two ancient civilizations; second, to integrate dissent into the second, and depict it only as its worst potential, an irregularity, a disease of culture; and, third, to advocate the reconciliation of the two forces, provided that the Hellenic remains healthy and reasonable. The title *Culture and Anarchy*, then, reflects the two possibilities and faces, the double potential of Hellenism, and the book outlines the benefits of its positive version for authority, the secularization of religion through/as art by culture. When uncontrolled, Arnold implies, culture may lead and turn into anarchy: today, for example, its unrestrained exercise threatens the foundations of old authority. Instead of letting the rivalry grow by fighting against culture, we should graft it onto authority; instead of rejecting and suppressing it, we should bring it to contribute to the creation of a new establishment. Because of its earthiness, inquisitiveness, and irreverence, the Hellenic is the real threat. It must be properly controlled by being fashioned after art and by being given the administration of science. Its role and territory, then, will be the construction of a national tradition and of a canon of perfection. It will thus serve and grace authority by making training mandatory and coercion redundant. Arnold's approach to the world of craft, custom, and festival is wholly aesthetic and allegorical, and intends to avert its politicization. The praise of culture and of the Hellenic advocates the former as the religion of the state and the latter as the supplementary beauty and supporting knowledge of the Hebraic. In his method of argumentation, there is no Hebraic without the Hellenic, and no defense of authority without both. To invoke the two is to distinguish the two and thus to hierarchize them, all in order to justify the rights and exercises of authority by the middle class. In all cases, the Hebraic is the modern – the modernist, the middle class, the moral, soul, progress, utopia, God; and the Hellenic is its different – its negative, its other, its supplement, its plenitude, its waste; its debauchery, its debasement, its debacle; its dis-interest, its dis-sent, its dis-sonance. The polarity reveals the aesthetic fashioning of man in the construction of bourgeois identity.

VI

Arnold kept the Hebraism–Hellenism distinction alive and graphic in all his work. But the most important return to it took place a few years after *Culture and Anarchy*, in *Literature and Dogma. An Essay Towards a Better Apprehension of the Bible* ([1873] 1892). The title of the former contains the two sides of the Hellenic; the new one contains both the Hellenic and the Hebraic. The first attempts to counter anarchy with culture, and combine the latter with authority; the latter tries to reconcile the two paradigmatic expressions of Hellenism and Hebraism, and show how they can work together. The opening sentence of its "Preface" sets the context unequivocally: "An inevitable revolution, of which we all recognise the beginnings and signs, but which has already spread, perhaps, farther than most of us think, is befalling the religion in which we have been brought up" (1892: v). Arnold worries about the "spread of scepticism" among the "*lapsed masses*" (1892: vi). People question the Bible and even reject it, while the churches can do nothing because theology is false and does not speak a relevant language. "Our mechanical and materializing theology, with its insane licence of affirmation about God, its insane licence of affirmation about a future state, is really the result of the poverty and inanition of our minds" (1892: xii). It excludes people by only confirming truth and power, without allowing them to participate through knowledge and understanding in faith. "Here, then, is the problem: to find, for the Bible, a basis in something which can be verified, instead of something which has to be assumed" (1892: ix–x). The old religious institutions and their practices cannot protect the faith from its corruption and the imminent collapse under the critique of dissent. A radical change is necessary, a broad revision. "The thing is, to recast religion. If this is done, the new religion will be the national one" (1892: x). A national religion is needed to unite all people – one respecting the rights and expectations of the middle class.

Arnold's aim is "to show that, when we come to put the right construction on the Bible, we give to the Bible a real experimental basis, and keep on this basis throughout" (1892: xi). His concern is the preservation of and correct approach to the Bible, so that a national religion may be built. And to this end he has again one remedy to recommend: "*culture*, the acquainting ourselves with the best that has been known and said in the world, and thus with the history of the human spirit" (ibid.). In regard to the question how can culture help our understanding of the Bible, this means "*getting the power, through reading, to estimate the proportion and relation in what we read*" (1892: xiv). Thus again he insists that culture, the secular religion, can help build the national religion. But, as we see, faith has been replaced by correct reading, believing by interpreting. Now the Bible is a text. His specific suggestion, as indicated in the title of the book, is that we read the Bible as literature, the Book as a book. We are not to take theological explanations for granted; we are rather to read and interpret. We do not necessarily need the mediation of the Church, since we can develop a direct, personal relationship with the text. Culture will help our interpretation, interpretation our religion, religion our nation. As Heinrich Heine would say, it is time to exercise our basic civil right, interpretation, and advance our collective interests.

It comes as a surprise to see literary reading invoked as an aid to understanding the Bible (and recasting religion). Only a century ago, it was biblical criticism that gave birth to literary criticism and gave it its first credibility by letting it use other texts as scriptures. But it is a sure sign of literary criticism's great success that it can now lend its power to revive the authority of the Bible. Arnold admits that the canon does not hold (1892: xxiv, xxvi). Reading-as-interpretation, which was generated by biblical studies, must now contribute to the study of the Bible, and return to its original model. This will be done through culture. In a major aphorism, Arnold states that "culture is *reading*" (1892: xxvii). The statement ought to be amplified: bourgeois culture is literary reading; the bourgeois religion of the state (=culture) is reading of the literary scriptures (=interpretation). Literary interpretation is the consummate experience of the bourgeois, his supreme and purest civil right: it is private, domestic, silent, passive, faithful. It is what culture teaches the middle class to do, how culture trains people in subjectivity, how it accommodates their libertarian claims in the privacy of beauty. That is why the topic of Arnold's book is "the relation of letters to religion," "their effect upon dogma," and "the consequences of this to religion" (1892: 5). Letters and religion, literature and dogma, culture and authority – we are back again in the realm of the secularized Hebraic.

Close to the end of his book, Arnold returns explicitly to the Hebraism–Hellenism dichotomy, using now the more specific terms "Greece" and "Israel." The former, he argues, gave to the world art and science, the latter conduct and righteousness. All these elements are important for a full life. "But conduct, plain matter as it is, is six-eighths of life, while art and science are only two-eighths. And this brilliant Greece perished for lack of attention enough to *conduct*; for want of conduct, steadiness, character" (1892: 320). Anxious to disperse any lingering misunderstandings of his earlier position, Arnold repeats himself in the "Conclusion" to explain that he never questioned the supremacy of Hebraism. Even when he praised culture, the importance of righteousness was paramount in his mind and taken for granted:

> And, certainly, if we had ever said that Hellenism was three-fourths of human life, a palinode, as well as an unmusical man may, we would sing. But we never said it. In praising culture, we have never denied that conduct, not culture, is three-fourths of human life. (1892: 345)

The awkward quantification highlights his despair about keeping the correct balance between the two forces. And his preference is strongly expressed, as when he states that "the revelation which rules the world, even now, is not Greece's revelation but Judaea's" (1892: 320). Of course it is true that the historical Israel perished too; but its lesson will never disappear.

> Thus, therefore, the ideal Israel for ever lives and prospers; and its city is the city whereto all nations and languages, after endless trials of everything else except conduct, after incessantly attempting to do without righteousness and failing, are slowly but surely gathered. (1892: 318)

The Hebraic moral revelation rules the world, and culture is called to serve its ultimate victory. It is also true that "conduct comes to have relations of a very close kind with culture" (1892: 345) and that it is "impaired by the want of science and culture" (1892: 347). But authority belongs to the rules of conduct, not the laws of art and science.

Culture, however, is necessary in its subservient role as a defense against skepticism and dissent that question authority. "And therefore, simple as the Bible and conduct are, still culture seems to be required for them, – required to prevent our mis-handling and sophisticating them" (1892: 348). In recognition of the fundamental law of our human being, which is both "aesthetic and intellective" and "moral" (1892: 349), we should accept that "even for apprehending this God of the Bible rightly and not wrongly, letters, which so many people now disparage, and what we call, in general, *culture*, seems to be necessary" (1892: 350). Aesthetics and literary criticism will repay a debt to theology and biblical interpretation, from which they arose, and will secularize their book, make it a literary text, translate it in terms the middle class understands and in situations it cherishes. Since reading/interpreting has become the fundamental experience, let us finally integrate it with its original subject, the Bible. Arnold seems to argue: if the bourgeois reads, give him the Bible to read, since you cannot expect him any longer to believe in it; now that he has no interest in the holy, give him the Bible for a truly disinterested experience.

In order to defend religion, the Bible, his class, and the nation, Arnold has to rehearse the Hebraism–Hellenism dichotomy. This thesis–antithesis rules his world, helps him make sense out of its divisions. It was also an integral part of his vision for the future. At the end of *Literature and Dogma*, he sees the possibility for a religious art that will only aspire to please God: "For, the clearer our conceptions in art and science become, the more they will assimilate themselves to the conceptions of duty in conduct, will become practically stringent like rules of conduct, and will invite the same sort of language in dealing with them" (1892: 349). But he does not stop here. He ventures into anthropological theories of race and, elaborating on the positions of Emil Burnouf, proceeds

> to talk about the Aryan genius, as to say, that the lore of *science*, and the energy and honesty in the pursuit of *science*, in the best of the Aryan races, do seem to correspond in a remarkable way to the love of *conduct*, and the energy and honesty in the pursuit of *conduct*, in the best of the Semitic. To treat science with the same kind of seriousness as conduct, does seem, therefore, to be a not impossible thing for the Aryan genius to come to. (1892: 349)

Here the Hebraism–Hellenism dichotomy culminates in a radical differentiation between the Aryan and the Semitic races, and their respective basic characteristics. The suggestion is one of conciliation, of possible combination of science and religion. Arnold always cared deeply about a natural religion that would express the race naturally, utilize the resources of culture (i.e., art and science), and unite all people in the community of a national state. His advice to his nation was to support and propagate the Hebraic religion with Hellenic art, authority and culture. From our historical perspective, it may seem strange to see him putting so much trust in the Aryan race. When he used his basic dichotomy to divide people in two races, so that the institutions of the new national reality could be justified, he was unable to predict the war of Aryan religion and culture that was to follow later. But by that time bourgeois authority was too fortified with culture to be stopped from barbarism. Apparently all that its early advocates, like Matthew Arnold, knew how to worry about was just anarchy.

★ I am grateful to Gregory Jusdanis and Michael Herzfeld for their suggestions on an earlier draft of this paper.

References

Arnold, Matthew (1892) *Literature and Dogma. An Essay Towards a Better Apprehension of the Bible* [1873]. London: Macmillan.

——— (1971) *Culture and Anarchy. An Essay in Political and Social Criticism* [1869]. Ed. Ian Gregor. Indianapolis/New York: Bobbs-Merrill.

Renan, Ernest (1868) *Questions Contemporaines*. Paris.

Part Three

Contemporary Culture: The Art of Politics

9

"Bringing it all back home"

American recyclings of the Vietnam War

John Carlos Rowe

> To put the old names to work, or even just to leave them in circulation, will always, of course, involve some risk: the risk of settling down or of regressing into the system that has been, or is in the process of being, deconstructed. To deny this risk would be to confirm it: it would be to see the signifier – in this case the name – as a merely circumstantial, conventional occurrence of the concept or as a concession without any specific effect. (Derrida, "Outwork," *Dissemination*)

That the "old name" of "Vietnam" has been put to work by American culture in ways that have kept it circulating, like some ceaselessly rubbed and fingered coin, needs hardly to be argued or demonstrated. No bibliography could cover the enormous representational productivity of this war: the war, as Michael Herr has written, covers *us* with an unparalleled volume and variety of texts, images, stories, studies, *products*. It continues to be *the* war of information, interpretation, and representation, even as such academic words must sound foolish and hollow to those who fought it. It is the most chronicled, documented, reported, filmed, taped, and – in all likelihood – narrated war in history, and for those very reasons, it would seem, the least subject to understanding or to any American consensus.

The war remains radically ambiguous, undecidable, and indeterminate to the American public, and the heterogeneity of what we name either "Vietnam" or the "Vietnam War" (the terms have become troublingly interchangeable in our cultural exchanges of late) seems to grow even more explicit and tangled with every new effort to "heal" the wounds, every new monument and parade, movie and book. Marxists have interpreted the war as a postmodern war, whose contradictions turned apparently stable ground into the quagmire secretly at the base of capitalist theories and practices. Post-structuralists have ingeniously argued that the war was conducted by the National Liberation Front by means of a postmodern military strategy that understood the hermeneutic possibilities of booby-traps and surprise attacks.

No western literary experimentalist could match the ingenuity of the Vietcong soldier who could turn a C-rations tin and American hand grenade into a lethal booby-trap or the terrible technology of the Bouncing Betty mine, which destroyed the lower limbs and genitals and attacked the myths of manhood and heroism of the American military.[1] Following, curiously, the same line of argument, conservative revisionary historians have claimed that the Vietnamese played both the Anti-War Movement in the US and the unfamiliarity and insecurity of the American troops in Vietnam as if they were syncopated parts of a carefully composed tune.

It is just half a step from the post-structuralist arguments concerning the ''uncanny'' bricolage of the guerrilla and the neo-conservative arguments regarding the NLF's ''exploitation'' of the ambiguities in the war to a familiar, old conclusion concerning the ''mystery'' of the Orient, with its thinly veiled master narrative of the cynical, amoral, wily, and manipulative Asiatic. Likewise, the arguments, such as Frances FitzGerald's, that insist upon the ''radical'' difference between Americans and Vietnamese as the ''source'' for the tragedy of the war reinforce popular conceptions of the strange, impenetrable ''East'' – an ethnocentrism reinforced by an American foreign policy that virtually censors news from postwar Vietnam. The familiar ethnocentrism and racism of imperialism are, then, reinforced, even reinvented with a difference, by way of claims for the radical ambiguity and undecidability of the Vietnam War, especially in its ''significance'' for American culture.

This is hardly a popular view, even among academics, for the convention has been that the government and other agents of ideology have been intent since Truman on ''stonewalling'' the issue of Vietnam, sticking to a ''story'' that the complex realities and historical events seem to name just that: a ''story'' to cover the bungling, ignorance, and human waste of the war. Thus the introduction of the critical perspectives: of feminists on the war, ranging from nurses' accounts of publicly unrecognized service in the field to feminists' deconstructions of the patriarchal myths informing our policies in that and other wars; of politically active veterans' groups, such as Winter Soldier and the Vietnam Veterans of America; of the Vietnamese refugee population in America; of scholarly accounts of the war and its representations. These and other ''perspectives'' seemingly have added a ''pluralism'' to the name of ''Vietnam'' that once was cynically imagined to be a fragile edifice constructed by the Executive and the Pentagon.

Linked to other, often subversive discourses – feminism, Marxism, post-structuralism, postmodernism – ''Vietnam'' would seem to have escaped the control of ideology and to have become the changeable referent for a process of complex cultural dissemination, whose counter-cultural force might be measured by the refusal of such a disseminated ''Vietnam'' to return to the Father, to return to its ''origins'' in American imperialism. Recycled and rechannelled, Vietnam would thus become one index of social change, of the sort that our failure in that war certainly required. In recent years, the standard and increasingly popular oral and epistolary histories of soldiers' experiences in and out of combat have been supplemented by such works as *Bloods: An Oral History of the Vietnam War by Black Veterans* (1984) and *Brothers: Black Soldiers in the Nam* (1982), oral histories in preparation of women's experiences in Vietnam, Lynda van Devanter (with Christopher Morgan), *Home before Morning: The Story of an Army Nurse* (1983), Patricia Walsh's fictionalized account of her work as a nurse in a Vietnamese civilian hospital, *Forever Sad the Hearts* (1982), and minority films of the "third-cinema" such as *Ashes and Embers*. These and other works have provided important alternative readings of the "Vietnam Experience," especially when they have been attentive to the specific ways gender, race, and class shaped the war experiences of the groups included. A common theme of these works, for example, has been the irony that exploited American minorities, in numbers disproportionate to their representation in our culture, were drafted to conduct a war of imperialist aggression against another exploited people – the Vietnamese. This is, of course, an old concern of liberal critiques of the war. In Peter Davis's documentary *Hearts and Minds* (1974/5), a native American veteran of the war is interviewed, dressed casually in jeans and work shirt, sitting on a rock above the desert in New Mexico: "In boot camp, I was called 'Ira Hayes' or 'Blanket Ass,''' he says, "depending on the drill sergeant's mood. I knew all about racism long before I went to Vietnam, yet I went anyway. From early on, I had been taught that the Marines were the best, the warrior class in this country, and I wanted to be a part of that." Van Devanter writes of the treatment of American nurses by US soldiers, who often mixed together sexist clichés about the easy morals of nurses with racist stereotypes applied to Vietnamese prostitutes. Accounts of the Presidio Riots in San Francisco and other racially motivated riots among troops in Vietnam have stressed the ways that the war foregrounded American racism as one of the motives for the war itself and well as one of the contributing factors in our military failure in Vietnam.[2]

In more general ways, the Vietnam Veterans of America and the Winter Soldier organization, which has conducted its own "trials" of American officials for "war crimes," represent the very significant political activism of veterans of the Vietnam war. Beyond these political organizations, there are little guerrilla theaters run by veterans in many urban centers and ensemble companies, such as the Vietnam Veterans Ensemble Theater Company in New York. Plays such as John de Fusco's *Tracers* (1980) have emerged from marginal theater groups to have fairly successful popular runs in traditional theaters. Written, directed, produced, staged, and acted by veterans, such plays reflect the political solidarity that many veterans consider the only means to redress such specific grievances as the medical consequences of the military's widespread use of the Agent Orange herbicide/defoliant in Vietnam and the psychological effects of PVS (Post-Vietnam

Syndrome) or PTSS (Post-Traumatic Stress Syndrome), as well as the more general consequences of America's foreign policy and the cultural repression of Vietnam as a socio-political entity. The work of such groups has focused significantly on symbolic, often experimental, forms, as part of larger political projects. Thus the New York Vietnam Veterans Memorial Commission collected and published *Dear America: Letters Home from Vietnam* (1985), as one part of their collective effort to shape the memory of Vietnam in the public conscience. Taken together, the memorial itself, covered with portions of letters and other documents from the Vietnam War, and *Dear America* constitute a single work of political art and activism.[3]

All of these counter-cultural or counter-hegemonic efforts on the part of marginal groups affected by the war can be said to be motivated by genuine commitments specifically at odds with the dominant ideology. Yet what is unique and singular about American ideology – *the* ideology first to have made the most efficient political uses of *diverse* cultural representations – is the speed with which it can incorporate a wide variety of critical perspectives in an enveloping rhetorical system designed to maintain traditional order and values.[4]

In 1985, CBS initiated a new project aimed at the home market: *The Vietnam War*, a series of videotapes culled from CBS television's vast files of war coverage. The overall organization of the series (three tapes have appeared thus far, with a fourth announced) and the internal organization of each tape are a confused mixture of thematic treatment and historical "record." The third tape, "The Tet Offensive," seems to suggest a purely historical approach, but the first two, "Courage under Fire" (concerned with the various means ground troops used to deal with the psychological pressures and physical dangers of an unpredictable war), and "Fire from the Sky" (about the air war), trace their themes in a loosely chronological manner. The curious combination of theme and history allows the editors of this series to make some interesting "revisions" of the war. In "Courage under Fire," for example, the selections proceeded from 1965 to 1971, except for the final and longest selection. Nearly half (30 of the tape's 61 minutes) of the entire tape is devoted to Charles Kuralt's famous "Christmas in Vietnam," a TV special filmed in 1965 and concerned with the daily lives and deaths in one military company along the DMZ. In this episode of the tape, whose length and formal organization make it stand out from the rest of the clips, the perspective follows that of Sergeant Bossulet, a handsome and extremely capable black officer, whose concern for his men's lives and his military duties are represented as nearly religious matters. The dramatic action focuses on one of Bossulet's very best men, José Duañez, a specialist in mines and booby-traps, who is from Guam, where he fought the Japanese during the occupation of Guam. In the hour-long tape, "Courage under Fire," we see absolutely *no* Vietnamese – no civilians, no ARVN, no NLF, no NVA – either living or dead. Even in a documentary as clearly ethnocentric as Eugene Jones's *A Face of War* (1968), Vietnamese villagers and a few dead VC suspects are shown. Duañez dies while looking for enemy mines, thanks to the clumsiness of Sgt Ray Floyd, a blustering, macho white soldier who insists that they "look for ourselves" for an elusive sniper, who has been attacking the camp daily from just beyond the tree-line.

My bare description already suggests some of the peculiarities of this 1965 mini-drama of Vietnam. The victimized Vietnamese people are erased from the entire tape, only to be replaced by an American soldier, represented as capable and admirable by his black commanding officer, who has our full sympathy: an American solider who is indistinguishable in appearance from a native Vietnamese, at least to western eyes. Duañez's death and Bossulet's anguished response to it conclude "Christmas in Vietnam," metonymically displacing the exploited Vietnamese and the political issue of American imperialism. Duañez's struggle with the Japanese on occupied Guam and the generally unquestioned heroism of American troops and people in the Second World War are invoked to shape our sympathies with an "Oriental."[5] That perception, however, is complexly mediated by our sympathies for Bossulet, who betrays no hint of racial self-consciousness as a black officer in a white man's war, and who epitomizes the ideals of a democratic leader intent on fulfilling his responsibilities to each man in his company. Even so, a quick interlude with Bossulet's family – his wife and five children – in California, trimming the Christmas tree, is accompanied by Charles Kuralt's explanation that Bossulet moved his family from New York, where he grew up, to Seaside, California, just before being shipped out, to keep his children from experiencing the urban violence in which he was raised.

It goes without saying that the ethnocentrism of American imperialism in south-east Asia and racism in the US are reinforced by such a narrative. Our admiration for Bossulet and Duañez is conditioned by their military discipline, as if to say that the fate of so many disadvantaged Americans – military service – is just what the advertisements say it is: an "opportunity" for these minorities to "lift themselves up by their own bootstraps." That these two "actors" happily perform these tasks of legitimating such institutions as the military and the bourgeois family and such myths as "self-reliance" and "leadership" is central to my claim that American ideology recognizes the dissemination of "Vietnam" into other critical discourses and employs their rhetoric for its own conservative purposes.

Even more interesting in this tape, "Courage under Fire," however, is the way that "history" is subtly revised along very similar lines – that is, by use of what have become counter-culture responses to the war in the past several years. The "historical" argument represented in this tape is constantly emphasized by Walter Cronkite's voiceover narration and occasional appearances, next to a map of Vietnam and an American flag, to provide historical "transitions" and explanations. But as I noted above, this "history" forms a loop, beginning with black-and-white combat footage from

1965, the year of our first troop landings at Da Nang, back to the 1965 "Christmas in Vietnam," filmed along the DMZ. Jones's documentary, *A Face of War*, although not released until 1968, was filmed during the 97 days he and his camera crew lived with the 3rd company of the 7th Marine battalion in 1966. In both films there is an air of innocence to the American soldiers which would be starkly absent, except for purposes of irony, from such later documentaries as Peter Davis's *Hearts and Minds* (1973) or even such establishment documentaries as *Vietnam: A Television History* (1983) and *Vietnam: The Ten Thousand Day War* (1984). The confusion of the soldiers, their fear, and yet their consistent sense of duty, unquestioned as yet by anti-war demonstrations back home or by the public exposure of government lies, are immediately evident in both "Christmas in Vietnam" and *A Face of War*. Like the American advisors in David Halberstam's *One Very Hot Day* (1965), these soldiers reflect the "equivocal realities," in one currently popular phrase, of the Vietnam experience.

To make such an argument in 1965 or 1966, as Halberstam and Jones do, is perhaps understandable as a consequence of the specific confusions of that historical moment. It is an entirely different matter for the CBS Video Library, however, to draw upon the ironies and confusion of 1965–6 in a 1985 "history" of the war. As I have argued, this "revisionary" history achieves its work not so much by returning us to a more "innocent" time – matching the cultural nostalgia that I think recurs with some regularity at moments of intolerable public anxiety and guilt – but more insidiously by *incorporating* those very counter-culture challenges that are aimed at bringing about searching re-examinations of social attitudes and public policies.

In the place of America's defeat in Vietnam, "Courage under Fire" concludes with this 1965 "episode" of war, which was filmed on the spot but carefully edited to construct the theatrical "narrative" that I have briefly described. Even before this obvious concluding displacement of defeat, however, the videotape does even subtler work of accommodation. Brief news clips and commentary by Cronkite refer to President Nixon's military incursions into Cambodia in April 1970. More than any other events during 1970–1, the US invasion of Cambodia brought public pressure on the Nixon administration to end the war. None of this is represented, of course, presumably on account of the "theme" of the tape – "Courage under Fire," which would preclude treatment of domestic affairs in the US. Yet, in even the most obviously conservative revisionary interpretations of this period in the history of the war, Nixon's reductions of troops in April (by 100,000) and November (another cut of 139,000), together with the nine-point North Vietnamese peace proposal to Henry Kissinger in June and Kissinger's revised peace plan in October, calling for the withdrawal of US forces within six months, are considered the most important events not only of this period but perhaps of the entire war.

In "Courage under Fire," however, Cronkite turns from the military incursions into Cambodia to the "courage of young American women in Vietnam." The year 1971 is thus fully occupied with interviews conducted with nurses in the 91st Hospital and with medics during a Medevac operation at Hawk Hill. These clips virtually "give evidence" that CBS news was concerned with the situation of women in Vietnam long before American feminists connected women in Vietnam with the more general issue of women and war. This cinematic "defensiveness" is, of course, some contemporary version of "protesting too much." Anyone watching the nightly news in the years 1970–1 is aware that "women in Vietnam" was a question of no consequence, an issue treated only to provide local color or "filler." Reread in light of feminists' challenges to patriarchal culture, in part by way of the exploitation of women in war (for many feminists *the* social phenomenon best suited to expose the irrationality and arbitrariness of patriarchal values), "young women in Vietnam" serve the CBS Video Library not just as the means of appearing *current* but to accommodate the very feminist challenge to the "cultural history" being rewritten by patriarchy itself. (It would be unfair, of course, to write *outside parentheses* of the many popular jokes and clichés that revolve around the mythologizing of "Walter Cronkite" as a kind of latterday, more benevolent "Uncle Sam" – a fatherly or avuncular figure presumably attractive to young women viewers for precisely these "virtues".)

It is, of course, no revelation that marginal, counter-cultural, critical discourses are "quoted" or "translated" in more popular forms of representation. The common conclusions about such borrowings, however, are that they function clumsily and obviously in the manner of what was once termed "radical chic" and that they thus betray even more readily the contradictions at work in that rhetoric of domination. When military fatigues and boots migrate from the soldiers to anti-war demonstrators to Vietnam veterans protesting the war to high-fashion models to punkers and, finally, to adolescents, the path of metonymic displacement is hardly direct or simple. Such semiotic "drift" involves complex combinatory (metaphoric) processes as well, which help prepare a market for such products as the new Saturday morning cartoon series, *Rambo*.[6]

The rediscovery of the military "hero" in such recent movies as *An Officer and a Gentleman*, *Top Gun*, *Rambo*, *Heartbreak Ridge*, and a host of other less publicized films is frequently judged to be merely a transparent recycling of familiar stereotypes out of Hollywood's mythopoeia in the 1940s and 1950s, as well as the various cultural sources for such "heroism" in the nineteenth century that informed such cinematic mythmaking.[7] The "motives" for this revival of the military hero are equally transparent, it would seem: Reaganite politics cynically manipulating the cultural mythology to condition the public support for new military ventures in the Caribbean, Middle East, and Central America.[8] Yet this ideological transparency is considerably more difficult to "see through" when we consider how centrally these heroes incorporate counter-culture and critical conventions.

Stallone's Rambo character incorporates counter-culture signs in just such a complex manner. *First Blood* (1982) opens with Rambo walking down a rural road somewhere in the north-west. Dressed in his army jacket and carrying a bedroll, his hair long, he looks like one of the many urban dropouts who moved to rural areas, especially in the north-west, in rebellion against urban, technological, military America. What he finds at the end of this road is the homestead of his black buddy in Vietnam, Delmar Berry. Built on the shores of a magnificent mountain lake, with a sublime backdrop of snowy peaks, the sturdy cabin, outbuildings, and signs of frontier life are designed to replicate an earlier, more authentic America. Delmar's wife is hanging wash and the children playing in the yard; she greets Rambo with suspicion. Delmar has died of cancer the previous summer, a victim of Agent Orange.

From this point on, the film will entangle a complex set of images to construct the character of Rambo: the NLF's guerrilla war for liberation, the American counter-culture's and anti-war movement's rejection of corporate America, the self-reliance of the early American settlers, the alienation of blacks, the modern western's anti-heroism, and the Vietnam veteran's exploitation during and after the war. The successful refunctioning of these figures, drawn as they are from radically different political, historical, and representational contexts, is achieved primarily by the overdetermined myth of American self-reliance and its conventional associations with isolation, moral conviction, nature, and resistance to the inevitable corruptions of civilization.

When he enters ''Holidayland,'' the small town where his troubles first begin, Rambo is identified not simply as a potentially troublesome drifter, but as part of the counter-culture. The sheriff's first words are: ''Wearing that flag on that jacket and looking the way you do, you're asking for trouble around here.'' In the police station, everyone in sight is wearing a patriotic American flag, so we must conclude that the sheriff takes Rambo's flag on his tattered military jacket to be a sign of defiance, a gesture of the anti-war movement. This is certainly confirmed by the sheriff's obsession to ''clean him up'' and give Rambo ''a haircut.'' All the derisive chatter in the police station, ''Talk about your sorry-looking humanity'' and ''Smells like an animal,'' recalls the clichéd red-neck contempt for ''Hippies'' that was so prevalent in the popular culture of the late 1960s and early 1970s. When they finally do force Rambo to take a bath, they hose him down with a large high-pressure fire-hose. The scene, with Rambo twisting and turning in the powerful spray, recalls scenes of civil demonstrations from Selma, Alabama, to Washington, DC, from the civil rights to the anti-war movements. Even as Rambo flashes back to scenes of his imprisonment and torture in Vietnam, the metaphors of American civil disobedience are associated with him. The brief images of Rambo's Vietnamese captors are clearly modeled – as they are even more explicitly in the sequel, *Rambo, First Blood, Part II* – after Hollywood stereotypes of Japanese captors in the Second World War. By the same token, in *First Blood* and *Rambo*, Rambo appropriates the guerrilla methods of the Vietcong for precisely those aims we would never grant them: freedom from despotic rule and irrational authority.

The "survival" equipment that Rambo takes with him into the north-west woods reminds us of his training in the Special Forces at the same time that it suggests "reducing life to its lowest terms." His cruel-looking, double-edged knife is, of course, a technological token, but it is a version of the Bowie knife and contains in its handle a compass and primitive needle and thread. The technology that invented this knife belongs to the inventiveness of the frontiersman, not the wizardry of the computer engineer. Dragging their sophisticated assault rifles through the woods, tripping over their bazookas, the police and National Guard who track Rambo epitomize alienated, post-industrial man. The targets of Rambo's furious violence are not primarily people, but the objects of technology: his survival skills outwit not only the National Guard troops who have him trapped in a cave but also the sophisticated weaponry they use to "seal" the cave's mouth.[9] The tools and weapons he employs in his rebellion, with the exception of his body and his knife, are taken from his antagonists, in perfect imitation of the Vietcong, whose booby traps and mines were often made of salvaged and captured American equipment, and in the tradition of the American frontiersman, that pre-eminent *bricoleur*.

Stumbling into the nightmare of "Holidayland," instead of finding the society he had sought with his friend, Delmar Berry, Rambo in *First Blood* gathers together a complex set of signs from high literature, popular culture, and film to figure the American desire for a purifying (and thus often violently cathartic) return to first principles and original revolutionary zeal. In Davis's *Hearts and Minds*, Americans dressed as Revolutionary War soldiers for a fourth of July celebration are interviewed concerning their views of the revolutionary struggle of the Vietnamese. One modern Minuteman reflects: "Yeah, I guess there's a kind of parallel there. They're fighting a foreign invader; they're fighting on their home soil. But the analogy stops about there. I mean, Oriental politics? You got to be kidding me!" Davis's argument that the Vietnam War was not a civil war between North and South, but a revolutionary war against foreign aggressors (French, Japanese, Americans) was a common one among anti-war activists in the period 1968–73. Rambo's "patriotism" is hardly a simple-minded Reaganism, although it certainly draws on that administration's blatant nationalism. By linking Rambo with the strategies and politics of the National Liberation Front, as well as with conventional images of an earlier, revolutionary America, Stallone transforms the victory of the North Vietnamese into a useful trope for American nationalism.

Thus when Rambo returns to Vietnam in *Rambo: First Blood, Part II* (1985), we are fully prepared for him to reject the authority of the US military in favor of his own revolution. Armed with the most sophisticated military hardware, Rambo is forced to cut himself free of most of this gear when the crew botches his parachute drop from a camouflaged corporate jet. As he cuts the cords binding him to the plane, Rambo is born again out of the necessity that has always fuelled his frontier spirit and powers of self-reliance. Reassured that his safety will be protected by an elaborate computer-command center, Rambo sententiously proclaims: "I've always believed that the mind is the best weapon." The sentiments are a bit shop-worn by now, but they are perfectly recognizable as derivative of the transcendentalist tradition.

"Man thinking" may be a strange motto for Rambo, but it is this claim that prevents his characteristic silence from representing the mere stupidity and brute physicality he represents for most of the Americans in the film (and watching the film as well). His meditativeness is frequently linked to his sympathies for the exploited and victimized. When he first meets the Vietnamese woman, Co, they greet each other in Vietnamese.[10] Both are dressed in clothing that resembles what American liked to call the "black pajamas" of the Vietcong. Co's first words in English are: "Did not expect a woman, no?" Yet, for all his macho, Rambo is no sexist; "We better go," is all he answers, as if to imply that the individual will always prove himself or herself according to actions, not the class, gender, or race with which he or she is associated. And when Co tells her "story," it is as simple as possible: "My father . . . killed; I take his place." Like Rambo, Co is liberated by necessity, in the burning fire of experience that dissolves national, class, racial, and gender distinctions. When Co acts most daringly and bravely, she dons the garb of a prostitute and plays the coquette with the Vietnamese guards to gain entrance to the POW camp. The "enemy" – American, Soviet, Vietnamese – views women as prostitutes, playthings for male desire. Thus when she opens fire on the camp guards and tosses Rambo his weapon, her violence is doubly, uncannily legitimate. As the conventional hero's "girl," she buys into tired sexist clichés about women who would die for love; as the sexually abused woman, she joins a host of feminine characters in recent films who turn their victimization by men into motives for violent revenge.

By associating Rambo with the frontier spirit, transcendentalist self-reliance, twentieth-century neoruralism, the anti-war and civil rights movements, the revolutionary zeal of the National Liberation Front, and the politicized Vietnam veteran, Stallone's two films incorporate these different and clearly critical discourses into the general revisionary view of the Vietnam War. In histories and novels, films and plays, the revisionary argument ranges from specific criticism of our military strategies in Vietnam to the more general indictment of our failure of national will. In *Rambo*, the

Soviets have arrived in Vietnam as quickly and pervasively as the most strident anti-communist from the McCarthy era had predicted. They have arrived for the same reasons the McCarthyites had claimed: the creeping immorality and weakness of the American people. The cynicism and expediency of the men conducting the ''Delta Force'' mission that Rambo joins are, with the exception of Colonel Trautman, Rambo's last link to the ''old Vietnam,'' born of the weakness that allowed us to accept defeat at the Paris peace talks. Even Trautman, of course, betrays his own name and reputation, caught as he is between the radical individualism of Rambo and the military bureaucracy he represents. In German, *traut* literally means ''dear'' or ''intimate,'' but *die Traute* used colloquially means ''guts'' or ''courage''; both senses are combined in the veteran's convention that his comrade-in-arms was his intimate just in so far as they shared the courage needed to survive the war. Even so, it is not anti-war activists, bleeding-heart liberals, weeping mothers, or lazy minorities who have brought about this corruption of American values in the semiotic code of the two films. Aligning himself with all of those, Rambo directs his rage against the military-industrial complex, thereby diverting genuine dissent into an affirmation of the very nationalism that took us to Vietnam in the first place.

The Vietnamese and the Soviets are unequivocal enemies, even if the Soviets express more admiration for Rambo as a man than any of the Americans. In a ''fair fight,'' Rambo can beat the most extravagant odds, as he does in his guerrilla war inside Vietnam, in an uncanny replay of the NLF's extraordinary military victory over the US. Colonel Trautman tells him before he leaves, ''Let technology do most of the work,'' and it is just this technology, together with its assumed manipulation of ''facts'' as information that can be used in a host of different contexts, that is Rambo's real enemy. Co's father ''worked for intelligence agency'' and was killed as a consequence; Co may take his place, but she does so as a warrior, rather than as a spy. *Rambo* marshalls counter-cultural arguments against high-technology America to claim that the war in Vietnam would have been better conducted by freedom fighters like Rambo and Co, both of whom represent the best virtues of their respective cultures. Although each of them seeks the ''quiet'' life, both are ready to protect the right of each individual to such a life by force of arms.[11] By claiming that our failure in Vietnam was caused by leaders far from the scenes of battle who relied blindly on technological representations of the war, *Rambo* diverts our attention from such obvious political issues as America's anti-communism since the Second World War and our commitment to outdated ''balance-of-power'' foreign policies. But this ''diversion'' is by no means simply achieved, in so far as it works through codes stolen from liberal critiques of the war and contemporary American culture.

The general distrust of representation in the two Rambo movies and many other popular works suggests an interesting twist on the *metaliterary* concerns of postmodern literature. Rambo's preternatural silence, his preference for actions over words, and the undeniable physicality of his body clearly set him apart from his commanders, who manipulate the data from their computers as deftly as Hawkes and Barth play the English language. Rhetorical flimflam belongs to technology-land, and in this regard *Rambo* aligns itself with those anti-war activists who judged our greatest immorality in the Vietnam War to have been the lies by which our leaders kept the public deluded for so long. Rambo is not too far removed from the laconic counter-culture characters in *Easy Rider* (1969), who clearly prefer immediate impressions to words and texts. Yet, this clear distinction between action and representation diverts us from the complex means the Rambo films employ to establish the character of "Rambo." Both films are full of scenes of data-gathering – rooms full of computers, telephones, and information-processing equipment that has the same status as hi-tech weaponry. These very sites are invariably the scenes of Rambo's greatest violent rage. He smashes the instruments of representation with the same zeal that the film-makers repress their own representational borrowings. The conventional scene of "literary self-consciousness" is transformed into an apocalyptic moment in which the machinery of technological man is repudiated, even though the viewer is perfectly aware that the cameras are still running. The consequence is a film, like *Rambo*, that appears unabashedly popular, nakedly trading on the easiest and most obvious sentiments, and thus appealing to a kind of crude stylization like that of folk-art (what the Germans aptly call *Naivkunst*). That "effect" generally has been quite convincing, I think, in so far as intellectuals have considered Stallone's two Vietnam movies to be mere symptoms of renewed and unreflective nationalism.

John Hellmann has argued persuasively that the virtual artistic silence on the subject of Vietnam from 1968 to 1974 was filled in part by the "Vietnam western or antiwestern, epitomized by Ralph Nelson's *Soldier Blue* (1970) and Arthur Penn's *Little Big Man* (1971)."[12] Bringing the war literally back home, especially by way of such equivocal mythic conventions as those associated with Manifest Destiny, served certain critical needs in those years, especially when it seemed so difficult to address the war without recourse to some historical explanation. That the anti-hero of Penn's film and Thomas Berger's novel finds his only genuine social attachment with the Cheyenne Indians seemed an unequivocal indictment of urban, technological America in 1971.

As Hellmann points out, "the Vietnam western was a short-lived genre," but the "war at home" narrative is one that has persisted and even flourished well into the 1980s.[13] The critical perspectives of such works seem invariably liberal, leftist, or otherwise minority-oriented, at least at the levels of design and intention, but they have also served some unintended ideological purposes that are part of the general appropriation and refunctioning of the name "Vietnam" that I have been describing in this essay. Above all, such works that displace the Vietnam War to America invariably turn a war of imperialist aggression into one of domestic conflict. Films like *Little Big Man* may qualify this generalization slightly, in so far as ''Vietnam westerns'' concerned with westward expansion do recall the white pioneers' origins in the more general project of European colonialism. Even so, this ''memory'' is quite faint in works dealing with nineteenth-century Manifest Destiny; few invoke the sort of imperialist zeal readily available in documents of that period.[14] And, of course, no ''Vietnam western'' could ironize completely the western conventions established by several decades of Hollywood productions aimed at glorifying the general heroics of ''How the West Was Won.''

Adventure narratives testing urban man against Nature have appeared in various forms and with considerable frequency in the past two decades. In the aftermath of Vietnam, several of these films, such as *Deliverance* and *Southern Comfort*, have used the American wilderness as an uncanny reflection of the jungles of Vietnam. Roy Hill's *Southern Comfort* (1983) develops this analogy in explicit ways, albeit without any overt reference to Vietnam. The convention of the belated rite of passage works with special irony in this film to suggest that those who missed the war still have all the motives and qualities to sink us in some other imperialist quagmire. National Guardsmen on weekend maneuvers in the Louisiana swamps run afoul of the secretive Cajuns, who have built their stills and shacks in the protective mazes of the bayou country. Although these weekend soldiers aren't supposed to carry live ammunition, one of them has brought along a couple of live rounds. Killing a Cajun more out of their own fear than the threat he poses, the guardsmen bury the body, burn the shack, and hope the bayou will providentially cover their tracks. Cajun society, it turns out, is considerably more tightly knit than these soldiers had expected, and a rather conventional revenge-plot develops as the dramatic means of exposing the ineptitude of these urban soldiers in the ways of Nature and man.

The conventional plot, with its frequent allusions to *Deliverance* and its use of virtually every popular cliché about Cajuns, bayous, and the Southern Gothic makes possible an extended meditation on a hypothetical war at home. The Cajuns are generally invisible, at best seen through mists or as hidden parts of the wilderness, especially as the camera follows the perspectives of the soldiers. Like the Vietcong, these backwoodsmen appear primarily as corpses or shadows. The guardsmen are particularly unsettled by the unfamiliarity of a region so close to home and their accidental discovery of the reality behind the myths of an American subculture. Picked off one by one by Cajun snipers and guerrillas, the soldiers begin to fight among themselves, replaying our popular notions of dissent among combat troops in Vietnam. When two surviving soldiers find their way out of the swamp and into a Cajun settlement, the analogy with Vietnam is made particularly explicit. A festival is underway, and Cajuns from the surrounding bayous arrive with cooked food and live animals as music and informal celebrations begin. Uncertain who their actual enemies are, the two soldiers are jumpy and paranoid, especially as the dialect of the villagers is incomprehensible to them.

The conventional narrative of the patrol in Vietnam generally ends in a village, where the troops try to overcome local, linguistic, and political differences to determine just who might be the enemy.[15] In *Southern Comfort*, the climactic scene in the Cajun village acts out some nightmarish reversal of the Vietnam patrol, since these two survivors are perfectly aware of how they have blundered into a foreign culture, whose appearances seem to deny the dangers they have just barely escaped. As it turns out, these two survivors, who are for that very reason cannier than their fellows in the ways of the woods as well as the ruses of foreigners, are not just paranoid; the "enemy" appears in the backroom of one of the shacks. Two "typical" red-necks silently attack them, and the two soldiers barely survive a graphically violent scene of self-defense. This attack unites the two soldiers in a camaraderie that for one urgent moment transcends their very different backgrounds and values, just as grunts in Vietnam are supposed to have found a special fraternity in battle.

The messages of the movie are explicit and unequivocal as far as the "Vietnam Experience" is concerned. Democratic America consists of many different sub-cultures, each of which has internal customs and practices that must be tolerated as part of a healthy cultural pluralism. By extension, our respect for this very American diversity ought to be applied in our foreign policies as well; we are, after all, a nation composed of most of the other peoples of the earth. The blunders of these National Guardsmen repeat the larger cultural errors of the government in the conduct of the Vietnam War. Frances FitzGerald's *Fire in the Lake* (1972) thus is the "proper" text for this film's reading of our involvement in Vietnam; our failure to understand the complex culture and history of the Vietnamese resulted in the repeated blunders we made in the name of "foreign aid."

Even so, *Southern Comfort* ironically turns against its own intentions or allows those intentions to be employed in the larger revisionary project of "winning" at last the Vietnam War. The Cajuns are Americans after all, which is why these weekend soldiers must learn to tolerate their different cultural values. As these two survivors hurry toward the helicopter LZ, repeating the common scene from the six o'clock news of the Vietnam War, they acknowledge the inevitable separateness of this sub-culture. "Understanding" has been primarily a matter of survival, not the means of cultural integration. Leaving the Cajuns well enough alone, the soldiers, like the American people after 1973, suggest that such foreigners want nothing more than their privacy. Yet, appearing in 1983, such a film appears at a moment when some critics of foreign policy were calling for American recognition of the Vietnamese government and thus the beginning of normal diplomatic relations that would lead to economic assistance.[16] Arguments in favor of economic assistance to Vietnam were often made in the context of reparations the US might reasonably be said to *owe* a country that it devastated economically, agriculturally, and politically. On the other hand, American journalists, like Stanley Karnow, returning to Vietnam in 1985 on the tenth anniversary of the fall of Saigon showed particular interest in reporting the demonstrable ''failures'' of the Vietnamese to rebuild their country.[17] Leaving Vietnam ''alone'' over the past ten years has certainly helped us shape a ''self-fulfilling prophecy'' of economic difficulty for that country. ''Understanding'' the Vietnamese for FitzGerald in *Fire in the Lake* involved our recognition that Vietnamese culture is fundamentally different from our own, an argument that in 1972 may have seemed to have been in the best interests of the Vietnamese, at least as far as the anti-war movement in this country was concerned. By 1985, however, it was clear enough that our deliberate *laissez-faire* foreign policy with respect to Vietnam could be justified on precisely FitzGerald's and *Southern Comfort*'s terms. The claim for the *radical otherness* of another culture belongs, as Edward Said has shown us, to the rhetoric of Orientalism; neglect or repression, even in the guise of ''tolerance'' of differences, is as much an imperialist policy as military intervention, as the economic misfortunes of contemporary Vietnam testify.

In another way, as well, *Southern Comfort* evokes Vietnam only to ''Orientalize'' it yet again. The film takes account of what we might term the phenomenology of the individual's ideological values, in so far as we *see* Cajuns only through the mediated visions of the National Guardsmen. Although these Guardsmen come from different classes and backgrounds, they are primarily from urban areas, and they share certain simple stereotypes of their antagonists. We see nothing *but* cultural clichés, then, such that the final ''appearance'' of the enemy takes the form of two vicious crackers such as haunt the motel dreams of northern travellers in the ''deep'' south. While dancing, talking, and eating go on undisturbed in the village square and houses, this ''back room'' is filled with the terrible unconscious of this sub-culture. In this scene of recognition, which is equally a classic scene of psychic *méconnaissance*, the enemy reminds us that this Cajun sub-culture maintains itself with homemade white lightning, petty theft, and the law of the rifle. The smiling women and helpful farmers look away when those more truly authorized attempt to apply their violent law to outsiders. Like the crackers who kill and sodomize the inept but finally ''decent'' suburbanites in Dickey's *Deliverance*, these Cajuns are perversions of natural law. The cathartic encounter with the Other is supposed to be purifying in *Southern Comfort*; it shows us otherness only to remind us of its terrors and thus allows us once again to estrange and repress it.

The violence of representation operates in the rhetorical uncanniness of American ideology, which can and does turn just about anything to its purposes with a sensitivity for the connotative and figural qualities of language unmatched by the subtlest artist, the cleverest critic. American ideology in the aftermath of the Vietnam War has managed to live up to Jean-François Lyotard's definition of the postmodern: "the presentation of the unpresentable."[18] What is *unpresentable* for American ideology is nothing as metaphysical as the Kantian Sublime that is the origin for Lyotard's postmodern; the *unpresentable* is precisely what is made so by those presentations by which a certain artificial *unconscious* is produced. The *unpresentable* is for America just what the play on words suggests: what cannot be made present because it exceeds the conventional boundaries of representation; what is for that very reason improper, *unpresentable* under the formal requirements of that system of representation. "Vietnam" as unpresentable? In a sense, that is, of course, the case for America in the 1980s; "Vietnam" remains the name that cannot be properly uttered, even though we continue to translate it into very presentable terms, very recognizable myths and discourses. What is finally *unpresentable* is the immorality of our conduct in Vietnam, and it is this unpresentable fact, which is hardly ambiguous and certainly unequivocal, that is made presentable in every sentence we speak of "Vietnam."

Lyotard's "postmodern" transcends the totalizing tendencies of conventional ideology; it embraces the variety, heterogeneity, and difference that we have come to associate with post-structuralist interpretation – not just a praxis or set of strategies, but a utopia, shaped by artists and polished by critics. There is something retrograde, naive in this "postmodern" culture, even if it is optimistic and properly committed to *justice*, the word never spoken properly, never made *presentable*, in our interpretation of "Vietnam." Jameson, in that formally curious "Foreword" he has written for the translation of Lyotard's *The Postmodern Condition*, makes it quite clear that Lyotard's postmodern seems merely to valorize the by-products of late capitalism: the play, becoming, and difference that may be the last resources of the *superficial* individual – the individual *flattened* by the very means of representational production that invented *him*.[19]

What lurks *beyond* modernism is by all means *postmodern*, but its knowledge is that everything that passed for "modernism" is always already serving other masters, those authors whose names never grace title-pages, but who are for that very reason the canniest of textualists. In this post-industrial society, the postmodern is our principal product: representations of the dizziest figurality, the most mobile features, capable of deployment at a moment's notice on the most distant shore.

The Fall premiere of *Miami Vice* in 1986 showed teenagers huddled around the television set, buzzing about Crockett's new Ferrari.[20] His black Daytona had been blown up by some black weapons smuggler, in a shocking but necessary demonstration of a hand-held rocket launcher. Word that a new car was on the way had been passed on the evening news. Stories in the

LA Times recounted Enzo Ferrari's threatened lawsuit over the use of the old car (it turned out to be a replica, originals being too expensive). At ten o'clock the prologue ran, but the disappointed teenagers watched documentary footage from Nicaragua spliced together with a docudrama of a cameraman and reporter recording American troops fighting in a small Nicaraguan village. At the close of this prologue, a Nicaraguan woman, carrying her dead child, walks into the camera's eye.

Miami Vice, surely the epitome of high-tech television cynicism and fashion for the past two years, had turned political. Between this prologue and Crockett and Tubs's desperate and futile efforts to get this videotaped "evidence" of America's military support of the *contras* on the major network news, the Ferrari appears to rev the interest of yawning teenagers. Most of the hour-long program, however, was devoted to Ira Stone, Crockett's old Nam buddy, who shot the tape in Nicaragua. Stone, who appeared first on the "Back in the World" episode of *Miami Vice* aired last Fall, is a cheap hustler, sometime photographer, and Vietnam vet, who uncovered their ex-CO's scam to recover and sell heroin smuggled into the US in body-bags.[21] You may remember that G. Gordon Liddy played Stone's and Crockett's former commanding officer in Vietnam. "Back in the World" played the conventions of the Vietnam War and the returning veteran as slickly and glibly as *Miami Vice* would have led us to expect. The plot of that episode was stolen out of Robert Stone's *Dog Soldiers* (1974), even Ira Stone's name seemingly borrowed casually from the author of that novel.

In the following Fall's episode, however, Liddy returned as the head of a paramilitary organization, funded by wealthy Nicaraguans living in Miami, that sends mercenaries to help the *contras*. Liddy makes it clear that their purpose is to make possible enough *contra* military success to convince Congress to provide more extensive and formal military assistance. The drama of the show revolves around problems of textuality. Crockett doesn't believe even his own eyes when he views Stone's tape, assuming that Stone must have "staged" the events as part of yet another confidence scheme. Convinced only after he has been beaten up by one of Liddy's henchmen, Crockett has to use the tape to bargain for Stone, whom Liddy has kidnapped. A black woman reporter for a major network – the only such reporter willing to take a chance on Stone's shady reputation – is killed and the tape erased (a powerful magnet is used for both tasks) by one of Liddy's men. In the end, Stone is killed, Liddy escapes, and Crockett and Tubs arrive too late as a transport plane full of mercenaries takes off for Nicaragua. Assuring the dying Stone that his tape will be aired on the six o'clock news, Crockett returns to his boat to listen to the news on the radio (fashionably counter-cultural, he doesn't own a TV), which reports events recorded on Stone's tape but claims them as new atrocities committed, not by the *contras*, but by the Sandinista government.

Culture critics must have watched this episode as a new phase in popular television. Until that time, *Miami Vice* had represented Latin Americans almost universally as canny, but brutal and brutish Bolivian (or Colombian or Venezuelan – such Latins being interchangeable) drug-dealers wearing woolen caps and riding in custom Mercedeses. This sudden shift from flagrant ethnocentrism and racism (Tubs's presence as a fashionable black vice officer previously seemed to permit the unlicensed representation of other blacks as either sexy women or utterly amoral criminals) to political self-consciousness worked back by way of Vietnam, linking our current adventurism in Nicaragua with our conduct in south-east Asia. Liddy, as both political personality and born-again actor, made that connection unavoidable. And the sympathies of the program with the Sandinistas and its opposition to American support for the *contras* were beyond any doubt, as was the program's studied self-consciousness regarding the political content of every popular medium from the evening news to *Miami Vice* itself.[22] The fact that the program did not quite represent Liddy's organization as an extension of the CIA seemed perfectly understandable and acceptable. *Miami Vice*, after all, could only go *so far* in such flagrant political commentary; direct indictment of the CIA would have led to a scandal that the network never would have risked.

Two days later, on October 5, a private American cargo plane, loaded with arms and munitions, is shot down in Nicaragua. An American from Wisconsin and veteran of Vietnam, Eugene Hasenfus, survives the crash and claims to be a mercenary working for an organization fronting for the CIA. Vice-president Bush, Secretary of Defense Schulz, and a host of other government officials deny any involvement in the affair, attributing it all to overzealous private citizens supporting the *contras*' cause. Read backwards, *Miami Vice* supports the government's claims. Some latterday G. Gordon Liddy, master of renegade dirty tricks, is undoubtedly behind this abortive venture. Just another Parable for Paranoics? Or is this crossing of popular television and foreign policy an ''event'' or ''coincidence'' that speaks the power of ideology to draw upon the slippery figures of its own representational machinery?[23]

At the Democratic National Convention in Chicago in 1968, anti-war activists staged a host of ''symbolic'' acts, claiming that their ''symbols'' were finally more powerful than the ''facts'' of the government's policies in Vietnam. In those innocent days, the playful, heterogeneous discourse of American Youth fought the referential claim of government experts. That was Modernism. Postmodernism begins with the discovery that the ''violence of representation,'' the ''unhinging'' of meaning, the freeplay of signification, and the *différance* that is the secret law of language are the principal means by which ideology – especially in America – maintains itself, always speaking in someone else's voice, never appearing as such, always more true to itself as it wanders from its apparent way: America, that incorrigible reviser of its elemental prose.

Notes

1 See Rappaport 1984.

2 Cortright 1975. Emile de Antonio's Marxist documentary, *In the Year of the Pig* (1969), focuses on the racism and ethnocentrism that motivated America's foreign policy in south-east Asia during the Vietnam War.

3 See my discussion of *Dear America* and the New York memorial in Rowe 1985–6.

4 In a Marxist analysis, the decentered, disseminated functions of American ideology would be perfectly consistent with the general hegemonic practices of capitalism, wherein the very "freedom" of the marketplace becomes the concept hiding the necessary classes established by "free-enterprise" economics – classes often as strictly maintained by hereditary transmission as the aristocratic ruling class in eighteenth-century England. In the case of post-industrial America, in which the principal products are representations, the generational transmission of power involves less investment capital than access to the popular media. Such access, often limited to a remarkably small number of people, a large percentage of whom come from media backgrounds or educations that stimulate such ties with those already in control, is often mystifyingly associated with "free expression" and other essential American liberties.

5 James (1985) discusses Eugene Jones's use of Hollywood conventions from Second World War movies in his documentary of Vietnam, *A Face of War*.

6 This series was introduced by a cartoon "mini-series," in which Rambo helps an avuncular Latin American leader to quell Communist insurgents in a small, Central American country. *Commando* (1986), a quick spinoff of *Rambo*, starring Arnold Schwarzenegger, makes the translation from Vietnam to Central America just as easily.

7 Applications to the various service academies increased dramatically in 1985–6; the press and officials at those service academies, especially the Naval Academy, attributed these increases to the favorable publicity given the military in such films as *Top Gun*.

8 Films like *Red Dawn* and Clint Eastwood's *Heartbreak Ridge* are obvious instances of Hollywood's continuing contributions to American foreign-policy-making. But that's an old story.

9 In several places in both *Rambo* films, implicit references are made to Huck Finn and Tom Sawyer, each of whom has certain curious affinities with Rambo. Rambo's struggle to escape the sealed cave does recall Tom's and Becky's escape from "Injun Joe's Cave" in *Tom Sawyer*, even though the analogy seems at first absurdly far-fetched. But Tom's escape from the cave *is* his special rite of passage, his means of proving his character above and beyond society. He "saves" Becky from the cave, when her father, Judge Thatcher, cannot. As a classic parable of sentimental manhood, the cave episode in *Tom Sawyer* prefigures Rambo's own special initiation.

10 Jones, *A Face of War*, shows one or two of the American Marines speaking Vietnamese with the villagers in "No-Name Village," which the Marines are trying to protect from NLF infiltration. But Jones shows these Marines from so many different angles and with such frequency in the scenes shot in this village that it appears that nearly the entire company speaks Vietnamese. The ways in which Jones gives us this impression in the documentary are classic instances of how documentary can be quite readily stylized from the most "factual" footage. It's worth adding here that the Vietnamese woman's name in *Rambo*, "Co," involves a pun on "C.O.," Commanding Officer. The Vietnamese name replaces

the American abbreviation, just as the liberated/liberating Vietnamese woman replaces Colonel Trautman as Rambo's leader.

11 Mark Lester's *Commando* (1985), an obvious spin-off of Stallone's *Rambo*, also invokes the virtues of the quiet, domestic family man to legitimate John's (Arnold Schwarzenegger's) violence. Like Rambo, John occupies a marginal position between military authority and the Latin American tyrant whose men kidnap his daughter. John aligns himself both with the traditional family and with minorities. Falling in love with a black woman who, like Rambo's Co, helps John overcome enormous odds, he returns to her with his daughter in the final scene of the film. When his former commanding officer, General Kirby, who arrives too late to be of much use in crushing the Latin American desperados, encourages him to return to his old commando unit, John tells him: "This was the last time." Kirby argues, "Until the next time," but John, holding his daughter and casting a cinematic glance at his waiting girlfriend – a glance that virtually constitutes the three as a family – answers decisively: "No chance." The family as a counter-force or alternative to corrupt and oppressive authorities is, of course, an old literary and cinematic convention. Even so, the special alignment of the family with the oppressed, exiled, and marginalized has assumed new significance in the representations of the Vietnam War. Many of the oral and epistolary histories, from Mark Baker's *Nam* and Al Santoli's *Everything We Had* to Bernhard Edelman's *Dear America*, encourage us to think that personal accounts of the war and family cohesion are remedies for governmental mismanagement. That these personal and domestic alternatives rarely address the political and sociological issues involved in the war ought, of course, to go without saying.

12 Hellmann 1986: 94–5.

13 ibid.: 95.

14 See, for example, Frank Norris's notorious, "A Neglected Epic" and "The Frontier Gone at Last," in Norris 1899. The following passage from "A Neglected Epic" is representative:

> Because we have done nothing to get at the truth about the West; because our best writers have turned to the old-country folk-lore and legends for their inspiration; because 'melancholy harlequins' strut in fringed leggings upon street-corners . . . we have come to believe that our West, our epic, was an affair of Indians, road-agents, and desperadoes, and have taken no account of the brave men who stood for law and justice and liberty, and for those great ideas died by the hundreds, unknown and unsung – died that the West might be subdued, that the last stage of the march should be accomplished, that the Anglo-Saxon should fulfill his destiny and complete the cycle of the world (282–3).

It's worth noting here that the American literary establishment, especially as it has been represented by twentieth-century Americanists concerned with the "uniqueness" of American nationalism, has helped to distinguish American colonialism from the evils of European ventures elsewhere around the globe. That such "nationalists," from Perry Miller to Sacvan Bercovitch, have stressed the New England origins of the nationalist myth is not incidental. By focusing on the religious persecution that drove the seventeenth-century Puritans to found the Bay Colony, such Americanists have helped to reinforce the traditional confusion of American colonialism with the revolution *against* it. On the bare evidence of John Endicott and John Eliot's accomplishments, however, the seventeenth-century American Puritans were quite eager to pursue familiar colonial policies of military domination and cultural hegemony in their little corner of the European empire.

15 The village scene of the conventional patrol often prepares for the cathartic firefight with the invisible enemy, now forced cinematically out of hiding, as in Ted Post's *Go Tell the Spartans* (1978) and Oliver Stone's more recent, but strangely anachronistic, *Platoon* (1986).

16 Among others, Noam Chomsky has made this argument in his many political writings on Vietnam, most recently in Chomsky 1985–6.

17 Karnow 1983: 42:

> The northern Vietnamese seem to be remarkably cheerful despite their grim poverty – perhaps because they are disciplined after a generation under Communism, and maybe because they never knew the affluence experienced by their Southern compatriots during the American era. In an ironic twist, however, the capitalistic propensities that the Communists were supposed to obliterate in the south are instead creeping northward with alarming speed. . . . The Communists cannot easily stop the trend, having been compelled by the economic crisis to loosen up in order to spur production. . . . Private entrepreneurs are . . . emerging in Hanoi, though more cautiously than in Ho Chi Minh City.

And so on.

18 Lyotard 1984: 78.

19 Jameson's "Foreword" to Lyotard 1984: xx, xviii.

20 *Miami Vice*, October 3 1986.

21 "Back in the World," episode of *Miami Vice*, Fall 1985.

22 When accused by Crockett of being an untrustworthy con-man, unlikely to be interested in political issues for any altruistic reasons, Stone counters: "You think you're much better? Driving fast cars, wearing fancy clothes, and making coke busts?"

23 This essay was first written before the scandal concerning the National Security Council's "Iranian Arms Deal" broke. One of the interesting albeit less-noticed facts that came to light during media coverage of that story was that the funds used to purchase the cargo plane carrying Hasenfus were obtained from the large sums resulting from the Iranian Arms Deal and laundered through Swiss banks for distribution to the Contras and various support groups. Popular opinion in the weeks following the disclosure of Poindexter's, North's, and other top administration officials' illegal activities had been that this scandal would cause the Reagan administration finally to give up its unpopular efforts to support the *contras*. Such popular opinions seem to suggest that the subtle rhetorical manipulations effected by ideology often end in such self-destructive scandals.

References

Chomsky, Noam (1986) "Reflections on the War in Indochina." *Cultural Critique* 3 (Spring): 10–43.

Cortright, David (1975) *Soldiers in Revolt: The American Military Today*. New York: Doubleday.

Hellmann, John (1986) *American Myth and the Legacy of Vietnam*. New York: Oxford University Press.

James, David (1985) "Presence of Discourse/Discourse of Presence: Representing Vietnam." *Wide Angle* 7 (4): 42–4.

Karnow, Stanley (1983) *Vietnam: A History*. New York: Viking.

Lyotard, Jean-François (1984) "Answering the Question: 'What is Postmodernism?'" in *The Postmodern Condition*. Minneapolis: University of Minnesota Press.

Norris, Frank (1899) *The Complete Works of Frank Norris*, 4. New York: P.F. Collier & Son.

Rappaport, Herman (1984) ''Vietnam and the Thousand Plateaus,'' in Fredric Jameson, Stanley Aronowitz *et al*. (eds) *The Sixties without Apology*. Minneapolis: University of Minnesota Press: 137–47.

Rowe, John Carlos (1986) ''Eye-Witness: Documentary Styles in the American Representations of Vietnam.'' *Cultural Critique* 3 (Spring): 126–50.

10

Figures of violence

Philologists, witches, and Stalinistas

Lucia Folena

I

> By the way, they should not be called Sandinistas – they've stolen the name of a true national leader who, in fact, rejected communism. What they really are, in truth are Stalinistas, because their revolution is a Stalinist one. (Ronald Reagan, 6/9/86 NC)[1]

After the dictator of Nicaragua, Anastasio Somoza, whose family had run the Central American country for forty-five years, was overthrown in July 1979 by the joint efforts of a coalition of forces known as the Sandinista movement, the Carter administration encouraged the revolutionary government politically and economically. Some twenty months later the new US administration led by Ronald Reagan cut off aid to Nicaragua and subsequently began to support groups of armed insurgents or *contras* who were declaredly attempting to topple the Sandinista government. This policy choice was motivated by the allegation that the country was becoming a Marxist state under the direct influence of Cuba and the Soviet Union, thereby endangering the security and independence of the other countries in the region, and indirectly threatening the US itself. From 1981 to 1987, despite the fact that no political change has been brought about in Nicaragua by the action of the *contras*, US political, financial, and military assistance to anti-Sandinista forces steadily increased, facilitated in its uninterrupted flow by the re-election of Reagan to the presidency in 1984.[2]

One is led to suspect that the six years of discussion over the hypothetical threats which the existence of the present Nicaraguan government posed to the US had in fact served a purpose quite different from that of achieving the overthrow of the Sandinistas. These six years offered a remarkable contribution to the construction of the political mythology of Reaganian culture; they provided the culture with a concrete, embodied figure of collective otherness as a semiotic device for its auto- and hetero-representation. Also, the production of the external enemy has often been a function of the need to mark an internal boundary, or the attempt to constitute the hegemony of this specific discourse of the Other in opposition to different discourses of otherness generated within the culture. The debate over Nicaragua and the Sandinistas thus went beyond the scope of presentday American foreign policy to offer an exemplary field of analysis of the ways in which restricted-code cultures use their interpretation of violence to displace and conceal the violence of their interpretation.[3]

Cultural representation as produced by Reagan's speeches and news conferences on Nicaragua was the outcome of a series of linguistic and rhetorical moves which are gestures of semiotic violence. Jointly, such moves bring otherness into existence in order to deny its legitimacy and to display, in the form of a moral imperative, the urgency of its obliteration. At the same time, and with further violence, the whole cultural construct of otherness is used to displace internal threats to the constitution of the speaking subject. In the Reaganian macrotext those threats often came from the Congress, whence Democrats repeatedly attempted to centralize their own more elaborated figures of otherness as constituents of a hegemonic language of the culture.

There are remarkable similarities between the violence of the Reaganian discourse of the Other and the rhetoric of self-production which shaped Jacobean culture through the language of witchcraft. Witchcraft is there merely on the landscape, the battleground, where a cultural war is fought; the witch provides the deixis or historical co-ordinates for the conflict. The real adversary is elsewhere; and I will try to discover where. More generally, the central function of the figure of the Other, in both Jacobean and Reaganian cultures, is that of bringing into existence a linguistic code for the production of the culture itself.

A culture depends on its ideology in order to exist; and since ideology as self-representation always and necessarily creates otherness, it can certainly be argued that ideology as such is intrinsically violent. But accusing a violence and simultaneously making it inevitable amounts to naturalizing and depoliticizing violence as a whole once more. What seems more important and profitable is identifying separate levels of violence, and showing that some of those levels are neither natural nor pre- or un-political. It is a matter of pluralizing violence, qualifying it, exploring the local and historical constitution of such levels. The violence I want to express is the violence of philology.

I am arbitrarily transferring the figure of philology from the domain of a specific, historically actualized methodology for reading and transmitting the written products of another culture to the domain of political interpretation in general, and subordinately to the domain of literary interpretation. The term here refers to any act of interpretation that is based on the assumption of the existence and knowability of a certain *intrinsic* property of objects of discourse in general which is named truth. The violence of philology consists precisely in its enacting a reconstructive project, in its proposing the possibility and the necessity of a restoration, presented as a reconformation to an origin(al). The violence of philology, that is, resides in its inventing the narrative or myth of a beginning in order to legitimize or even naturalize some kind of intervention on an object constituted as other. Furthermore, the violence of philology bypasses this Other by using it as a figure to speak the Same.

Philology is one of the complex strategies that an ideology may use in order to define the culture and its margins. While not every ideology adopts philological strategies to assert itself, all philology is ideological, so much more so when it lays claim to universality and apoliticalness; when it pretends – as it invariably does – to transcend the limits of the here and now in its quest for an ultimate truth. One of the most striking manifestations of the mystification of philology resides in the fact that the philologist does not represent "itself" as an agent of change, but creates for itself the image of the catalyzer of a reassuring return or homecoming, thus incessantly repeating the gesture which substitutes the innocence of a mythical past for the potential threat and disorder of a non-representable future; according to Reagan, in Nicaragua "it is a case of returning to the original goals of the revolution" (4/15/85: 418). The *contras* are simultaneously attempting to reconstruct a lost origin and to re-establish a coherence between the present state of the Sandinistas and their anterior being:

> if the Sandinista regime remained, but remained true to the original purpose of the revolution – this is a government by one faction of the original revolution. . . . So what is being struggled for there is a restoration of the original revolution (Reagan 7/21/83 NC: 1,032).

The circularity of the imagined development has a satisfying symmetrical quality. That this original revolution can never be said to have existed, except as recounted from a postlapsarian state of severance from it, is evidently of no relevance. The past is only a figure of the desire of the present.

The Other is only a figure in the discourse of the Same: shaping the Other becomes a cultural priority. Naming or renaming, defining (marking linguistic boundaries), translating Sandinistas into Stalinists and *contras* into freedom fighters, amounts to waging the first offensive in the war of cultural representation. Far from being the preliminary gesture of a further, and more real, struggle it is this liminal move that exhausts the virulence of the battle. For any subsequent play of truth and counter-truth is substantially the empty, however gory, re-enactment of a violence that has *already* taken place with the institution of the primary boundary between Same and Other:

> Sympathy with the *contras* . . . is more and more pervasive. In fact, the peasants now call them *Los Muchachos*, the affectionate term they once used exclusively for the Sandinistas. And what do they now call the Sandinistas? Well, the latest workers' chant is "the Sandinistas and Somoza are the same thing." (4/15/86: 448)

Blood is thus shed in the name of a name, battles are lost and won over the establishment of a universalizable designation.

II

In 1591 a case of witchcraft, which initially seemed to involve mere threats to the orderly course of urban life, was brought to light in the Scottish town of Trenent, where "one *Dauid Seaton* . . . being deputie Bailiffe in the saide Towne, had a maide seruant called *Geillis Duncane*, who . . . [suddenly] took in hand to help all such as were troubled or greeued with any kinde of sicknes or infirmitie," thereby arousing Seaton's suspicions "that she did not those things by naturall and lawfull wayes, but . . . by some extraordinary and vnlawfull meanes" (*Newes from Scotland* 1924: 8–9). Her confession gradually uncovered a general conspiracy against King James's life. Other documents state that the conspiracy was led by an aristocrat, the King's cousin, Francis Stewart, Earl of Bothwell.[4] The narrative has thus already received an added layer of interpretation, a metaphorical reading of the Devil. The ground, moreover, has shifted from the broadly social to the strictly political, and from lower-class disorder to upper-class treason. It is my contention that *Newes from Scotland* should instead be read literally. The Devil, I am arguing, *was* the real enemy.

He appeared more than once "in the habit or likenes of a man" to a group of 200 witches who had sailed by sea in sieves to the town of North Berwick; on such occasions he "did greatly inveighe against the King of Scotland," and when asked "why he did beare such hatred to the King . . . answered, by reason the King is the greatest enemy he hath in the worlde" (*NFS* 1924: 14–15). James, who was taking part in the examinations of the accused, initially refused to believe their stories and "saide they were all extreame lyars," until one of the witches managed to convince him by repeating to him "the verye woordes which passed betweene the Kings Maiestie and his Queene at Vpslo in Norway the first night of their mariage," which made him give "the more credit to the rest" (*NFS* 1924: 15).

Thus, the language of witchcraft had not yet been completely assimilated by the monarchy as late as 1590. The executions in which the episodes recounted in *Newes from Scotland* resulted in fact opened the first great Scottish witch-hunt, which was to last seven years. It was as if the monarchy was then being offered a new global discourse of otherness through which to constitute itself. The offer came primarily from the Presbyterian Church. This might partly account for the king's initial resistance to the assimilation of the language, whose political usefulness was not yet completely self-evident on the one hand, and whose adoption by the monarchy might on the other hand have imported the risk of a total surrender to the theocratic premises of Presbyterian theologians like Andrew Melvill.[5] If the main enemy of the culture was one and the same with the enemy of God, then the culture must needs identify itself with the self-proclaimed figures of God within it. The battle was then to be fought on the ground of control over the central representation of God within the culture. *Newes from Scotland* is thus the account of a spectacular sleight of hand – the centralization of the monarchic trope and the consequent displacement and marginalization of the Kirk: "the King is the greatest enemy he hath in the worlde." The monarchy was here constituting itself as the subject of the discourse of the culture through appropriating the Other of the Church as its own Other, and consequently appropriating the Church in its own order.

It is therefore necessary to contextualize the shift between social lower-class and political upper-class witchcraft, or rather the synecdochic reinscription of the former in the general framework of the latter. The reinscription served the secondary purpose of asserting control over the nobility and its centrifugal tendencies; the main battle, however, was being fought on the ground of the culture's identification and on that of the control over the means of its self-representation. In this battle the aristocrat was, like the witch, rather a place or deictic element than a direct object of discourse. If the devil is eventually translated into a nobleman, this nobleman never acquires the status of a complete self-included sign but remains, essentially, a figure of the devil – a vehicle for the centralization of the monarchy in the culture by means of a plausible *antitheton* to it. Despite any imaginable clash between king and nobles, it would make no sense at all for a monarchy to constitute itself through designating the aristocracy as its absolute Other.

A number of moves have been made to enact this final victory in the battle of representation. The setting of the narrative has changed from individual and private, and lower-class, to collective and political and upper-class witchcraft; from transgression or (etymologically) boundary-violation to treason; from inversion to subversion. Even this, though, is only part of the picture. Treason, subversion, and aristocracy are taken back to the stage of religion because only there can the complete assimilation of monarchy and culture be celebrated. If the politics of self-representation in a restricted-code culture is always and necessarily displaced into the ethics of self-legitimation, and if the language of ethics in Jacobean culture is one and the same with the language of Christian religion, the constitution of any form of alterity other than one directly associated with the negative side of religious postulates would there require a double displacement. For such an endeavor would demand the negative ethical investment of an ethical neutral object as a preliminary phase in the process of its alienation. To become the absolute Other of a restricted-code culture the New World savage, for example, would need to have already been made *antonomastically* recognizable as bad or ignoble. The witch in Jacobean culture is, by virtue of her demonic alliance: she is thus its ideal figure of otherness, its perfect *antitheton*.

That represented in *Newes from Scotland* is a central moment in the history of social difference: all witchcraft, no matter whether high or low, individual or collective, is for the first time reduced to one general paradigm of otherness, thus singularized and rewritten as a global aggression against the institutions of the Scottish monarchy and *consequently* Kirk. Seven years later witchcraft is officially identified in the *Basilikon Doron* as one of the central threats to the constitution of the culture: ''there . . . [are] some horrible crimes that yee are bound in conscience never to forgive: such as Witchcraft, willful murther, Incest (especially within the degrees of consanguinitie) Sodomy, Poysoning, and false coyne'' (James VI/I 1682: 23). James's 1597 *Daemonologie* functions as a perfect counterpart for the *Basilikon Doron* – a treatise of the Other to complete the general theoretical assessment of the monarchy provided in the latter work. Read together – and in association with a number of other texts such as the *Trve Law of Free Monarchy* – *Daemonologie* and *Basilikon Doron* in this sense function as a sort of manifesto of Jacobean culture; for, as Stuart Clark argues convincingly, ''Both in genesis and in content the *Daemonologie* may be read as a statement about ideal monarchy . . . [and] there is a sense in which demonism was, logically speaking, one of the presuppositions of the metaphysics of order on which James's political ideas ultimately rested'' (Clark 1977: 156–7). It is certainly not by accident that the *Daemonologie* was republished twice in London in the same year in which James Stuart ascended the English throne, and translated into Dutch and Latin in 1604, and into Latin again in 1607 (Robbins 1981: 277).[6] This was the first time a monarch explicitly addressed the issue except as a cultural problem in a treatise.

The most important feature of the *Daemonologie* is then not the content of the document but its very existence, as a clear manifestation of the urgency on the monarch's part to take a definitive stand in the discussion on the existence of witches. By 1597 witchcraft has been completely singularized and externalized as a general conspiracy to destroy the very foundations of the culture; if ''meanes'' and ''waies'' are ''diuerse,'' all associations with the Devil tend

> to one end: To wit, the enlargeing of Sathans tyrannie, and crossing of the propagation of the Kingdome of CHRIST, so farre as lyeth in the possibilitie, either of the one or other sorte [i.e., of Magicians or Witches], or of the Deuill their Master. (James VI/I 1924: 34)

The word ''tyrannie'' is significant here, for it manifests an absolute political symmetry between the culture and its adversary, which are seen as perfect reversed mirror images of each other. The same antithesis receives a good deal of discussion in the *Basilikon Doron*:

> consider . . . the true difference betwixt a lawfull good King, and an usurping Tyrant, and ye shall the more easily understand your dutie herein: for *contraria juxta se posita magis elucescunt*. The one acknowledgeth himselfe ordained for his people, having recieved from God a burthen of

government whereof he must bee countable; the other thinketh his people ordained for him, a pray for his passions and inordinate appetites, as the fruites of his magnanimitie. And therefore, as their ends are directlie contrarie, so are their whole actions, as meanes, whereby they preasse to attaine to their ends (James VI/I 1682: 18)

If monarchy is authorized by God, it is also a synecdoche or worldly actualization of this order, a "true pattern of Divinity," since this "forme of Government, as resembling the Divinity, approcheth nearest to perfection . . . Vnity being the perfection of all things" (James VI/I 1642: 3). In the same way, it is very clear that – given the restriction of the code – every tyranny ultimately goes back to Satan, as in an automatic reciprocation of the situation described in the passage of the *Daemonologie* quoted above.

Likewise, the political implications of a Biblical simile used in the *Daemonologie*, namely "*disobedience is as the sinne of Witch-craft*" (James VI/I 1924: 5), are made clearer by the inclusion of lack of respect for parents and princes among the unforgivable crimes listed in the *Basilikon Doron* together with the sin of witchcraft (James VI/I 1682: 24). Here is how the culture views the association:

It is a principle of the Law of nature . . . that the traytor who is an enemie to the State, and rebelleth against his lawfull Prince, should be put to death; now the most notorious traytor and rebell that can be, is the Witch. For she renounceth God himselfe, the King of Kings, she leaves the societie of his Church and people, she bindeth herself in league with the devil . . . (William Perkins, *Discourse of the Damned Art of Witchcraft*, 1616–18, quoted in Clark 1977: 176)

James's is no simile at all, but a complete identification.

The Other in the text of the Same speaks the language of carnival and inversion.[7] Satanic rites offer a perfect reversed mirror image of the ecclesiastical and secular rites of the monarchy; contrary to what still happens with certain previous lower-class inversions of the order of the state,[8] however, no playfulness and no levity are implied but a grim determination to subvert the whole organization of society:

the deuill as Gods Ape, counterfeites in his seruantes this servuice & forme of adoration, that God prescribed and made his seruantes to practise. For as the seruantes of GOD, publicklie vses to conveene for seruing of him, so makes he them in great numbers to conveene . . . for his seruice. As none conueenes to the adoration and worshipping of God, except they be marked with his seale, the Sacrament of *Baptisme*; so none serues Sathan, and conueenes to the adoring of him, that are not marked with . . . [his] marke (James VI/I 1924: 35–6)

The *Daemonologie* here goes into further detail in the establishment of a perfect correspondence between each element of the Christian ritual and each feature of Satanic worship. What seems very clear is that the whole description of the evil practices of the witch serves the fundamental purpose of offering a displaced reaffirmation of legitimate power.[9] Antithesis ultimately gives way to antiphrasis, a trope associated with irony as a figure of opposition and inversion, and with allegory as a figure of displacement.

James's culture thus recycles the medieval and Renaissance language of inversion and tropes it to constitute itself. That language is no longer used to appropriate and neutralize, through the explicit authorization – or the tolerance – of spaces of disorder within the order of the culture, potential lower-class threats are converted to the latter. It has now acquired the main function of providing the culture with the materials for a differential self-representation. The language of festival or sports is simultaneously being removed from a dangerous semantic contiguity with the domain of inversion, and authorized precisely to the extent that it lays no claims to reproducing a negative symmetrical image of the discourse of legitimate power. That of the "world turned upside down" is no longer a tolerable fiction of displaced lower-class violence, for inversion is gradually becoming synonymous with subversion, and the language of carnival will soon be appropriated by the radical discourse of Puritan revolution: "Freedom is the man that will turn the world upside down" (Winstanley 1941: 316; cf. Hill 1972).

III

The witch presupposes the philologist. Every production of otherness requires a science of the Other, whether restricted – as in James's discussion – to a representation of "*genus*, leauing *species* and *differentia* to be comprehended therein" (1924: xiii), or articulated into extensive taxonomies of alterity, as in a number of other demonologies. The language of this science has a more than superficial similarity with that of the science of language. If speaking the Other amounts to speaking the Same, and speaking the Same means speaking the language, or culture, of the Same, in the two senses of "using" that language and of "pronouncing (ideological) statements about" that language, or culture, then there is a definite affinity between philology and production of cultural otherness; they are both metalinguistic practices of the culture.

James's redefinition of witchcraft in the *Daemonologie* takes the literal form of a philological endeavor to establish coherence, at several separate levels, between names and realities, "wordes and deeds," present manifestations and their origins. Etymology therefore becomes an essential factor in the self-authorization of the discourse of the culture, whether it is used to validate or to invalidate the labels traditionally applied to the Other:

This worde *Magie* in the Persian toung, importes as muche as

> to be ane contemplator or Interpretour of Divine and heavenlie sciences: which being first vsed amongs the Chaldees, through their ignorance of the true divinitie, was esteemed and reputed amongst them, as a principall vertue: And therefore, was named vnjustlie with an honorable stile. . . . And this word *Necromancie* is a Greek word . . . [meaning] the Prophecie by the dead. The last name is given, to this black & vnlawfull science by the figure *Synedoche* (James VI/I 1924: 8–9)

In classical rhetoric, etymology is one of the two forms of definition *par excellence*, together with "usage" or *consuetudo*, to which it opposes its quest for an "original truth" based on the "transparency of the original coincidence between word-form and word-meaning" (Lausberg 1960, 1: 255). Over the descriptive gesture of a definition by usage etymology imposes the prescriptive requirements of a return to the origin which makes *consuetudo* tantamount to corruption or degeneration: "*Magie* . . . was named vnjustlie with an honorable style"; "they should not be called Sandinistas." This is exactly, on the level of the word, what philology as critical and political strategy does on the level of discourse and text and culture.

The Other is a complex intertext of previous representations of cultural difference. Once produced in its Jacobean form, this intertext in turn acquires the power of generating a global project of intertextual revisionism. Thus a number of Biblical passages, many of them directly mentioned in the *Daemonologie*, are rewritten to accommodate the new enemy of the culture: "Thou shalt not suffer a witch to live" (Exodus 22: 18). Jacobean philology then also manifests itself in the form of a need for the establishment of the greatest possible coherence between one discourse and another, among those which are fundamental in constituting the culture; almost an attempt to create a complete and consistent encyclopedia of the monarchy. Instead of placing an interpretative diaphragm between pre-existing texts and itself, the culture materially interpolates those texts with its interpretations to such an extent that it eventually produces those interpretations in place of, or as if they were, the texts themselves. Every culture of course does that, in a figurative sense; Jacobean culture does it literally, precisely when it is at its most philological. For a central paradox of philology, generally speaking, consists in its displaying a reading of the origin or truth of an object as if it were the origin or truth of that object.

Philology presupposes that reality and truth are separable from the entity to which they pertain, and that the latter points at them as at something that is located behind or beyond it. Implicit in philology is thus an urge to traverse the object in order to attain its truth. The postulation of the primacy of truth and the consequent dematerialization or disembodiment of the object represent, after the institution of restricted-code otherness, the second specific level of the violence of political philology. If the object is the cultural Other this disembodiment can, and does in Stuart Scotland, also take the extreme form of judicial torture.

For judicial torture is also a way of traversing the body and denying its materiality – making it into a mere sign of a truth that resides beyond it. The tortured body is an object of the violence of interpretation, a mere text to be read by the hermeneutics of inquisition. The body is thus just a means of access to its own real meaning, and simultaneously an obstacle, an opaque diaphragm interposed between the interpreter and the full disclosure of that meaning. Reading the body must needs amount to destroying it in order to substitute for an empty signifier the plenitude of inquisitorial signifiedness. Judicial torture in the Renaissance and in the seventeenth century is constituted as an interpretative practice, and as such requires a science and a language: far from being a mere expression of uncivilized brutality, of violence as disorder and unpremeditation, it manifests the civilized, code-structured cruelty that characterizes the violence of the culture.[10]

The North Berwick witches were subjected to grievous and repeated torture; most of them refused to confess, despite that, until the mark of Satan was found on their bodies. The man who had been accused of being their ''Regester'' or ''Clarke'' and who ''did take their othes for their true seruice to the Diuell, and . . . wrot for them such matters as the Diuell still pleased to commaund him,'' a ''Doctor *Fian*, *alias Iohn Cunningham*,'' managed to escape from prison after a first extorted confession. When he was caught he retracted, ''notwithstanding that his confession appeareth remaining in records vnder his owne hand writing, and the same therevnto fixed in the presence of the Kings maiestie and sundrie of his Councell.'' A search for a new satanic mark on his body produced no result, so that ''for more tryall of him to make him confesse, hee was commaunded to haue a most straunge torment'' whereby ''His nailes vpon all his fingers wer riuen and pulled off with an instrument called in Scottish a *Turkas*, which in England wee call a payre of pincers'' (*NFS* 1924: 18–27). This time he persisted in his denial, which convinced the accusers that ''the deuill [had] entered into his heart'' and that he must therefore be executed.

Torture is thus not primarily meant to produce a confession, but to reproduce a truth which has already been uttered as pertaining to the accused's body and does not even demand the echo of ''his'' avowal in order to manifest itself fully. Torture is still necessary as a self-validating strategy of the inquisitor: for it provides this further evidence of guilt that consists in the absence of the confessing voice of the Other. Thus the material body is excluded from language and deprived of its voice. Even the confession, however, is a mere echo of the question.[11] A metaphysics of anteriority makes perfect sense of what would appear a logical aberration from the vantage-point of a different culture – the fact that the voice generated by torture is not, cannot be, the voice the body identified as its own before the torture. The body thus becomes the sounding board of a word that is already there, since it belongs to its speaker, the inquisitor, but needs a material obstacle to manifest itself. The Other has been reduced to the pure function of providing a definition, an outer margin for the self-voicing of the culture.

Philology separates literature from politics. It is certainly important to overcome the separation by asserting that literature is political; but it is also necessary to stress that political discourse is literary, inasmuch as it is a form of cultural representation. Therefore, it seems not only legitimate but essential from an anti-philological perspective to read contemporary political discourse as literature, by way of complementing the political reading of literature. For denying the separation between literature and politics in an earlier culture at times amounts to producing an equally arbitrary division between the political discourse of the past, whose distance from the critic makes it a plausible element of a revisionist canon, and the political discourse of the present, which stronger marks of involvement with the real world prevent from attaining the completeness of textuality. If King James's *Daemonologie* is literature – that is, representation – so are Ronald Reagan's speeches; and they should be read as such.

It is the disconnection of the literary from the political that has allowed the philological strategies at work in the latter to go unnoticed. Philology erects barriers but simultaneously endeavors to distract attention from itself and its operation. Retracing the effaced steps of philology in the field of political utterance amounts to uncovering the double violence of disconnection and obliteration.

Contemporary philology covers its traces by staging continual displacements. The way in which Ronald Reagan speaks the figure, or is spoken by it, is typical of this kind of strategic evasiveness. The terms of the debate, despite their appearing only too clear-cut and glaringly well-defined, persistently elude the interlocutors, who are thus forced to map out the erratic orbits of these unattainable objects and consequently kept from redirecting their attention on the semantic constitution of the terms themselves. By this operation discourse is simultaneously translated into metadiscourse, that is, into unmediated ideology, and prevented from exposing its translatedness:

> Q. Mr. President, theologians recently criticized you for saying, in defending the military budget, that the Scriptures are on our side. I wonder, do you think it's appropriate to use the Bible in defending a political argument?
>
> THE PRESIDENT. Well, I was actually speaking to some clergymen, and I checked that with a few theologians – if it was appropriate and – well, what I meant about appropriate, was I interpreting it correctly? . . . they seemed to think it was perfectly fitting, yes. It was a caution to those people in our own country who would, if given the opportunity, unilaterally disarm us. . . . I've found that the Bible contains an answer

> to just about everything and every problem that confronts us, and I wonder sometimes why we don't recognize that one book could solve a lot of problems for us. (Reagan 2/21/85 NC: 213)

The answer displaces the question: "was I interpreting it correctly?" The legitimacy of producing such a practically-oriented political interpretation is not even at issue in the answer, as if it were a given. It all boils down to finding the right interpretation, one authorized by all experts. What takes place here has much in common with what happens to the context of a pun. For in fact the displacement or transcoding of the question on the part of the answer is effected by means of the rhetorical figure of antanaclasis – the dialogic repetition of the same word in a different sense. The term "appropriate" in the question interrogates political deontology, and therefore directly involves the ethical code; the answer makes the word raise a strictly philological and hermeneutical issue of coherence between interpretation and true meaning, thus shifting the discussion to the epistemological ground.

In Nicaragua "the struggle . . . is not right versus left; it is right versus wrong" (Reagan 3/1/85: 245). Reaganian philology uses the violence of the pun in the same way in which Jacobean philology uses the violence of etymology – to displace politics into ethics and provide the gestures of the culture with a metaphysical foundation. Reaganian philology reformulates the right–left opposition into an up–down antithesis, "up to the ultimate in individual freedom . . . or down through statism until you arrive at totalitarianism" (7/16/84: 1,031). This literalizes a displacement of syntagm into paradigm, and politics into ethics, which governs the whole production of this rhetoric of otherness. Shifting the ground from political categories to moral postulates amounts to making the Inside into a necessary and natural manifestation of the Positive, since "right" and "wrong" are implicitly constructed as preceding any imaginable cultural distinction.

Such a postulation of anteriority inevitably results in singularizing its recipient by making it into a universal – an archetypal, transhistorical, and transcultural category which can then be used to define other objects. Thus when Ronald Reagan speaks of a "truth . . . revealed to man in his soul" (6/20/86: 845) as a basis for political action he invokes the anteriority of "truth" over "soul" and "man" in order to universalize the last term. The violence of philology then also resides in the reduction of plural to singular which takes place on both the pole of the object and that of the subject of discourse as a result of a series of synecdochic moves. The singular subject produced by humanism – man or culture – can then become the repository of a grammar of values and truths that renders itself automatically available for being reproduced, unchanged, on the pole of the singularized object.

For the third level of the violence of philology, after the definition and the postulation of truth, stages the universalization of the subject's ideology, or return to truth. This is a more strictly political gesture, because it presupposes a direct and concrete intervention on the subject or Other. If the ultimate truth about "man" is "his" human-ness, or the universality and invariance of some characteristics of human nature, no matter how defined, then this constant must also exist in the Other, as the aspect which determines the Other's fundamental relatability to the Same:

> the new day will come when U.S.-Cuban relations, based on the ideals of democracy, will be restored. The philosophical foundation of these relations already exists in the hearts of the Cuban people. . . . It is this principled devotion to truth – not the so-called truth of the Communist Party but the real truth as revealed to man in his soul – that will be the source of that new day and a new era of peace between our nations. (6/20/86: 844–5)

The Other is like the Same in that they are both human, and unlike the Same in that it acts differently. Now, if the truth about the Same is identical with the truth about the Other, the right behavior consistent with that truth must needs also be one. The Same cannot evidently presuppose that its own behavior is inconsistent with its fictions or postulates; therefore it must assume that the behavior of the Other is wrong. It then becomes a moral imperative for the Same to reduce the Other to the right behavior, that is, to restore a coherence between the Other's pristine meaning or truth and its present form or manifestation: "as long as the people of Nicaragua are still striving for the goals of the revolution that they themselves fought, I think we're obligated to try and lend them a hand" (4/1/85: 397). The search for coherence takes place at both poles of the discourse of political philology: "supporting the *contras* is not only legal, it's totally consistent with our history" (2/16/85: 187).

IV

The common narrative of violence is the story of something which, in contrast to "power," does not have "the form of the rule" (Foucault 1980: 92), and can therefore *by definition* only pertain to the Outside of a culture. This is, in particular, the Reaganian narrative of violence: "America's relations with Cuba will improve on the day that Cuba stops exporting violence . . . [and] terror, . . . stops facilitating the traffic of drugs . . . [and] begin[s] to respect the rules of international life" which are epitomized "by the Golden Rule: 'Do unto others as you would have them do unto you'" (6/20/86: 843). The antithesis between "rule" and "violence" in fact universalizes a specific rule, and the culture that produces it, into the only acceptable condition of existence: "the rule of law must supplant violence" (3/10/83: 380). The lack of "rule" and "norms of civilized behavior" manifests the lack of the codes and boundaries of culture; violence is absence of culture. No culture could narrate itself as a whole in terms of absence of culture, because, even if a synecdochic identification between *the* culture and culture as a transhistorical construct did not take place, the very language used to tell this tale would inductively produce a linguistic order and a rule to belie the tale. Violence thus must needs pertain to the Other, and the Other be the enemy: "The greatest threat [to liberty in 1986] comes from the self-proclaimed enemies of our freedoms and civilization," those who "are wedded to violent means of obtaining power"; the group includes "Communists and other radical forces" (6/23/86: 883). The narrative of the violence of the Same, conversely, is told by the Same as a story of power, or even force. Both terms are plausible, at times acclaimable, qualifiers of the Inside. The antithesis of power and violence is precisely the figure through which Reagan establishes a difference between the October 1983 military intervention in Grenada and what he elsewhere refers to as "the new colonialism of the totalitarian left" (2/22/83: 276): "Let us always remember the crucial distinction between the legitimate use of force for liberation versus totalitarian aggression for conquest" (7/19/84: 1,044).

If violence is displaced onto the Other, it is because otherness and violence, from the point of view of the Inside, are one and the same. An attribution of violence is in the Reaganian code a sufficient condition for the production of cultural difference. The word "violence" takes on different connotations depending on the term which is set in a relationship of opposition to it as well as on the object against which the violence is directed. Violence is thus both a state or condition of the Other and a product of the Other's boundary transgression. In the first sense it is antithetical to "freedom," or "democracy," and synonymous with "tyranny," "empire," "dictatorship," "communism," "totalitarianism," and "statism." In the second occurrence it is opposed to "rule," "order," "civilization," or "legitimate . . . force," and identical with "disorder," "aggression," "subversion," "terror," "cruelty," "repression," and "oppression." A further distinction can be established in this case between the violence that is exercised from outside against the primary boundary between Same and Other and the violence which appears to take place outside that boundary. The terms of "disorder," "aggression," and "subversion" apply to the former, those of "cruelty," "repression," and "oppression" to the latter; "terror" can refer to either one.[12]

The existence and operation of the first three levels of the violence of philology – definition, truth-postulation, and universalization of the subject's truth – are contingent upon an externalization of the central term, violence; this gesture, moreover, serves the complementary purpose of distracting attention from itself and the violence which calls it into being. In other terms, cultural self-representation in a restricted-code situation depends on the representation of the violence of the Other, and this representation in turn functions as a *pharmakos*, or a material exorcistic device for the displacement and concealment of the violence of the Inside.

Philology is a figure of time, in that it constructs fictions of anteriority; it creates an illusionary past in order to bring about changes in the present and simultaneously deny their occurrence. Reaganian philology repeatedly translates the time of its myth of origin into a space of metonymic difference, or physical antithesis. The Reaganian Other occupies a *literal* Outside of the culture, located beyond an actual border which is thus invested with a full load of symbolic connotations. Recurrently, the otherness of Nicaragua is produced in terms of a concern with the integrity of geographical boundaries: "for the first time in memory we face real dangers on our own borders . . . we must protect the safety and security of our people. We must not permit dictators to ram communism down the throats of one Central American country after another" (7/19/83: 1,019). The southern border of the US embodies the difference between Same and Other: "This Communist subversion poses the threat that a hundred million people from Panama to the open border of our South could come under the control of pro-Soviet regimes" (5/9/84 AN: 677).

The Other is constantly threatening to break through, and break down, the boundaries/definitions of the Same. Any attempt against those boundaries is an act of violence; conversely, the violence of the Outside invariably manifests itself as a violation of the physical or cultural boundaries of the Inside. For if the violence of the Other resides in its lack of moral, legal, and linguistic codes, or in the inversion/perversion of those codes, it is not very surprising that the threat posed by the Sandinistas should in fact have a double quality (at least), as violation of borders and violation of pledges, and that the former should function as a material counterpart of the latter. The Sandinistas have "violated" both their "promise" not to "attempt to overthrow any other government in Central America" (4/14/83: 546) and their "contract . . . with the Organization of American States . . . that if they had the help of the OAS in persuading the Somoza government to step down and let their revolution take over" they would respect "freedom of press . . . and . . . all the freedoms that we enjoy here in this country" (7/21/83: 1,031). The text of the past with its breaches of faith and legitimacy is actually a troped version – a *figure* – of the text of the future, "the long-term danger to American interests posed by a Communist Nicaragua, backed by the Soviet Union, and dedicated – in the words of its own leaders – to a 'revolution without borders'" (6/24/86 AN: 868). Because present and future events constantly repeat the past, the past in turn can be used to generate symmetrical extrapolations about the future. The transhistorical constitution of the subject, on either side of the boundary, results in dehistoricizing history, or excluding the ultimate possibility of its producing actual change. Both Same and Other are always "consistent" with their respective histories, and they will always be, until the Same succeeds in establishing that final and universal coherence which is brought about by the obliteration of the Other, by its "return" to the origin.

Reaganian philology needs to establish another distinction in order to bring out the essential qualitative difference between intra- and intercultural delimitations, since only the latter are fundamental. The former must therefore be renamed; ''seams'' map out the place of the Same, ''boundaries'' mark its impassable limits:

> Well, I went down there [to Central America] and asked them [?] what we should do. I said, ''How can we recognize that we are, even when we cross a border into another country here in this hemisphere, we're among Americans.'' And I said, ''Maybe you think that we've taken that name for ourselves because we call ourselves Americans. But,'' I said, ''that's just because of the name we picked for our country. We can't walk around calling ourselves a united state.'' But they were very pleased to hear that we felt that way and recognized they, too, were Americans.
>
> And the other day . . . the . . . chairman of one of those [Hispanic] business groups . . . gave me a line that I think we should all adopt. He said, ''We talk about the international boundaries here in the Americas. Well,'' he said, ''a boundary kind of sounds like something that separates us.'' He said, ''Why don't we start calling them the international seams, because seams bring you together and hold you together.'' (Reagan 8/26/83: 1,181)

This is a typical case – and a very literal one at that – of what I have been calling universalization or singularization. The coincidence of the name of the continent with its synecdochic appropriation on the part of the largest and most powerful country on the continent (''Americans'' are ''US citizens'') is here exploited to prove the necessity and positiveness of an association of all American countries. This association in turn functions as a reversal of the original synecdoche (''Mexicans,'' or ''Brazilians,'' for example, are ''Americans'') through an intermediate metonymic stage (''The US'' and ''Mexico'' are physically contiguous). The same strategy is at work here as in the singularizing politics of the confrontation with the Other. Sharing a definition – that of being Americans; that, also, of being jointly delimited by the primary boundary – presupposes sharing, along with it, a system of postulates, and consequently a specific behavior: ''Here are . . . some 600 million people . . . all worshipping the *same* God. While we may speak different languages, we all have the *same* pioneer heritage. We came here from other countries in search of freedom'' (1/21/83: 89; my italics); ''We [in the US] want the *same* thing the people of Central America want'' (3/10/83: 377; my italics). For there is a ''natural bond between all Americans'' (5/8/84: 666), whose ''birthright'' and ''heritage'' it is ''to live in freedom'' and whose duty in order to achieve this is to ''stand together'' (2/20/86: 249). Every attempt to rewrite the definition, to stress the differences marked by what the Reaganian discourse of the Other encodes as non-essential boundaries, or ''seams,'' receives the same reductionist answer; some borders are there only to lessen distinctions.

The primary boundary itself is subjected to two different interpretations, depending on whether this side of it is in view or that. Its outer part is much more imposing and forbidding, being a wall:

> today democracy in Latin America constitutes a swelling and life-giving tide. It can still flow its powerful, cleansing way into Nicaragua – the Communist wall against it is high, but not yet too high – if only the House takes action [to approve *contra* funding]. (6/16/86: 771–2)

This is of course the same wall that the repressive Other has built "to keep people in" (6/30/83:966). The Other's wall delimits an inverted Eden, invested as it is with the double function of preventing access, or "holding" the tide of democracy "out" (8/5/86: 1,050), and precluding egress. The erection of the primary boundary is thus attributed to the Outside, in another of those characteristic displacements which are specifically aimed at producing the moral aberration of the separating gesture.

In order to make the universalizing gesture possible, another dissymmetry has to be postulated. The violence exercised beyond the boundary against "people without politics" (4/15/85: 450) actually has the Same as its object. The government of Nicaragua is severed from the people of Nicaragua and produced as an utterly external entity, an absolute Outside created by "a faraway totalitarian power" (8/12/86: 1,124) which is "trying to establish a beachhead of tyranny, subversion, and terror on the continent of the Americas" (6/16/86: 841). The *contras*, conversely, are completely internal to the country, since theirs is the "largest peasant army raised in Latin America in more than 50 years" (6/24/86: 867); and they are "the people of Nicaragua" (2/11/85: 171). In order to take the outsideness and illegitimation of the Sandinistas to its furthest possible point, "Nicaragua" as a whole has to represented as "a country held captive by a clique of deeply committed Communists at war with God and man from their very first days" (3/14/86: 325). A "rescue mission" aimed at liberating the synecdoche of the Same walled in by the Other would appear as a logical consequence, although "the freedom movements in Nicaragua and Angola are so strong that, if it weren't for the intervention of Cuban and other foreign forces, they could win all by themselves" (6/20/86: 843).[13] Any help to the *contras* would thus not represent a real external interference, but a mere response to such an interference on the other side.

Yet "the question" as far as the US is concerned "is not only about the freedom of Nicaragua . . . but who we are as a people" (6/24/86: 868). As happened in Jacobean times, the discourse of the Other is ultimately the discourse of the Same. "The issue is human freedom, and it towers above all partisan concerns" (6/16/86: 772). Foreign policy is a figure of domestic politics. For there is an internal Other. There are those who "are confused" and "refuse to understand" (8/13/83: 1,127); they are "the new isolationists," and "would yield to the temptation to do nothing" to restrain "the aggressor's appetite'' (6/9/84: 681). The real enemy in this sense is the Congress: ''the opponents [in a vote on *contra* aid] in the Congress of ours, who have opposed our trying to continue helping those people, they really are voting to have a totalitarian Marxist–Leninist government in America, and there's no way for them to disguise it'' (4/29/85: 557); ''Congress must understand that the American people support the struggle for democracy in Central America'' (2/17/85: 187). The Congress is thus within the boundary what the Sandinistas are outside of it, an external disturbance in the solidarity of the administration with ''the American people.'' The violence of philology comes full circle. After instituting the definition, dematerializing its object to attain the truth about it, and producing the necessity of a return to that truth, it enacts this ultimate bypassing of the Other which consists in using it as a sounding board to voice its own anxieties about cultural hegemony.

Notes

1 All quotations labeled ''Reagan'' refer to the *Weekly Compilation*; abbreviations should be read as follows: NC, News Conference; AN, Address to the Nation.

2 It is necessary to state that what is at issue here is absolutely not the *personal* ideology or culture of Ronald Reagan. Ronald Reagan is here a positionally representative voice coming from the political center of a culture which, very schematically speaking, proposes itself as extremely compact, easily decodable in its postulates, and self-produced as offering a theoretically inexhaustible set of binary oppositions (between itself and all that does not pertain to it) that are ultimately reducible to one fundamental antithesis – that between ''us'' and ''them.''

3 A restricted-code culture sees itself as the positive pole of an antithesis whose negative term is the Other, and organizes its whole ideology in terms of sets of antitheses which are ultimately reducible to a fundamental dichotomy (e.g., that between good and evil). Stuart Clark provides a genealogy of what I am calling code-restriction in western culture from early Greek religion to the late Renaissance (1980: 104–10).

4 Cf. Rosen 1969: 190; Clark 1977: 158–9. In fact the only reliable contemporary authority for the plausibility of an identification of Bothwell with the North Berwick devil seems to have been James Melvill; according to his memoirs the devil himself and some of the witches had ''blessed'' a wax image with the words, ''This is king James VI, ordained to be consumed at the instance of a nobleman, Francis, Earl of Bothwell'' (quoted in Bingham 1979: 130; cf. Clark 1977: 158). The identification became almost universally accepted only after the publication in 1921 of Margaret Murray's influential book, *The Witch-Cult in Western Europe* (Cohn 1975: 107; Hoyt 1981: 85–6).

For the Gowrie plot, which presented another case of aristocratic witchcraft/treason at James VI's court in 1600, see Mullaney 1980: 32–47.

5 ''The serious challenge to the authority of James VI came not from the nobility but from the Kirk, and principally from the extremist Presbyterian party led by Andrew Melvill'' (Bingham 1979: 137). In 1596 Melvill addressed James with these words: ''there are two kings and two kingdoms in Scotland. There is Christ Jesus the King, and His Kingdom the Kirk; whose subject King James the Sixth is, and of whose kingdom not a king, nor a lord, nor a head, but a member'' (quoted in Bingham 1979: 137; cf. Willson 1967: 38).

6 The republication of the *Daemonologie* in London in 1603 had a specific function. It was no longer a matter of producing the ideology of the monarch as against ecclesiastical and other threats to its absoluteness, but of reproducing it. The Church of England as institution was by definition one and the same thing with the monarchy, and, moreover, it was interested in prosecuting exorcists rather than witches (Clark 1977: 163; Greenblatt 1986).
7 Cf. Bakhtin 1968, and recent rediscussions of the Bakhtinian category of carnival such as Stallybrass and White 1986, esp. 1–43. On inversion, besides Stallybrass and White 1986: 6–26, see Tennenhouse 1986: 153. On the relationship between the language of carnival and witchcraft in western culture see Clark 1980: 102; Stallybrass 1982; Serres 1968: 219–31.
8 Those, for example, which took place in Rabelaisian France (Bakhtin 1968); or those which were tolerated on the Elizabethan stage in the 1590s (Tennenhouse 1986: 40–3).
9 In general, as Mary Douglas observes, "witchcraft beliefs are essentially a means of clarifying and affirming social definition"; for "witchcraft sharpens definition where roles are ill defined" (Douglas 1970: xxv, xxx). Peter Stallybrass stresses that such beliefs "are less a reflection of a real 'evil' than a social construction from which we learn more about the accuser than the accused" and that "Witchcraft accusations are a way of reaffirming a particular order against outsiders, or of attacking an internal rival, or of attacking 'deviance'" (1982: 190). This is also the thesis expressed by Stuart Clark with reference to both James and the wider context of the discourses of witchcraft in general (cf. Clark 1977; Clark 1980).
10 See Foucault 1979: 33–69.
11 The answers of tortured witches as recorded in trial papers are often literal repetitions of the inquisitors' questions or consist in a single word, *affirmat*, "she says yes." Cf. Robbins 1981: 502.
12 In fact the separation between the two latter types of violence in the Reaganian macrotext is only a provisional gesture and does not presuppose a reproduction of the "natural" or metaphysical difference between Outside and Inside.
13 A "rescue mission" or "commando" subtext governed the intervention in Grenada, which was presented as motivated by the need to free a number of US medical students, and subordinately the whole population of the island (e.g., 6/9/86: 842), from "captivity" under a Marxist regime: "this was not an invasion . . . [but] something in the nature of a commando operation, and it was a rescue mission" (12/24/83: 1,742). Both US students and Grenadians were thus represented as figures of the Same. The actual presence of US citizens in the regions of the Other is therefore not indispensable to justify military interventions, since "captive" people are ideally located on the same side of the boundary where both "God and man" reside, and their physical restoration in this site is a moral imperative.

References

Bakhtin, Mikhail (1968) *Rabelais and His World*. Trans. H. Iswolsky. Cambridge, Mass.: MIT Press.

Bingham, Caroline (1979) *James VI of Scotland*. London: Weidenfeld & Nicolson.

Clark, Stuart (1977) "King James's *Daemonologie*: Witchcraft and Kingship," in Sydney Anglo (ed.) *The Damned Art: Essays in the Literature of Witchcraft*. London: Routledge & Kegan Paul: 156–81.

——— (1980) "Inversion, Misrule and the Meaning of Witchcraft." *Past and Present* 87: 98–127.

Cohn, Norman (1975) *Europe's Inner Demons: An Enquiry Inspired by the Great Witch-Hunt*. New York: Basic.
Douglas, Mary (ed.) (1970) *Witchcraft Confessions and Accusations*. London: Tavistock.
Foucault, Michel (1979) *Discipline and Punish: The Birth of the Prison*. Trans. A. Sheridan. New York: Vintage.
——— (1980) *The History of Sexuality, I*. Trans. R. Hurley. New York: Vintage.
Greenblatt, Stephen (1986) "Loudun and London." *Critical Inquiry* 12: 326–46.
Hill, Christopher (1972) *The World Turned Upside Down: Radical Ideas During the English Revolution*. New York: Viking.
Hoyt, Charles A. (1981) *Witchcraft*. Carbondale-Edwardsville: Southern Illinois University Press.
James VI/I (1642) [1598] *The Trve Law of Free Monarchy, Or, The Reciprocall and Mutuall Duty Betwixt a Free King and His Naturall Subjects*. London, printed by . . . T.P.
——— (1682) [1598]*Basilikon Doron, Or, King James's Instrvctions to His Dearest Sonne, Henry the Prince*. London, printed by M. Flesher for Joseph Hindmarsh.
——— (1924) [1597] *Daemonologie, In Forme of a Dialogue*. London: Bodley Head.
——— (1967) [1611] *The English Bible Translated out of the Original Tongues by the Commandment of King James the First*. New York: AMS.
Lausberg, Heinrich (1960) *Handbuch der literarischen Rhetorik. Eine Grundlegung der Literaturwissenschaft*. 2 vols. Munich: Max Hüber Verlag.
Mullaney, Steven (1980) "Lying Like Truth: Riddle, Representation and Treason in Renaissance England." *English Literary History* 47 (1): 32–47.
Newes from Scotland (1924) [1591] *Newes from Scotland, Declaring the Life and Death of Doctor Fian. . . . With the True Examinations of the Saide Doctor and Witches. . . .* London: Bodley Head.
Reagan, Ronald (1983) *Weekly Compilation of Presidential Documents*, vol. 19, nos 1–52. Washington, DC: Office of the Federal Register.
——— (1984) *Weekly Compilation of Presidential Documents*, vol. 20, nos 1–52. Washington, DC: Office of the Federal Register.
——— (1985) *Weekly Compilation of Presidential Documents*, vol. 21, nos 1–52. Washington, DC: Office of the Federal Register.
——— (1986) *Weekly Compilation of Presidential Documents*, vol. 22, nos 1–52. Washington, DC: Office of the Federal Register.
Robbins, Rossell Hope (1981) *The Encyclopedia of Witchcraft and Demonology*. New York: Bonanza.
Rosen, Barbara (ed.) (1972) *Witchcraft*. New York, Taplinger.
Serres, Michel (1968) *Hermès ou la communication*. Paris: Minuit.
Stallybrass, Peter (1982) "*Macbeth* and Witchcraft," in J.R. Brown (ed.) *Focus on Macbeth*. London: Routledge & Kegan Paul: 189–209.
Stallybrass, Peter and White, Allon (1986) *The Politics and Poetics of Trangression*. Ithaca, NY: Cornell University Press.
Tennenhouse, Leonard (1986) *Power on Display: The Politics of Shakespeare's Genres*. New York–London: Methuen.
Willson, David H. (1967) *King James VI and I*. New York: Oxford University Press.
Winstanley, Gerrard (1941) *The Works of Gerrard Winstanley*. Ed. G.H. Sabine. Ithaca, NY: Cornell University Press.

11

The violence of rhetoric

Considerations on representation and gender

Teresa de Lauretis

For Umberto Eco

> Older women are more skeptical in their heart of hearts than any man; they believe in the superficiality of existence as in its essence, and all virtue and profundity is to them merely a way to cover up this 'truth', a very welcome veil over a *pudendum* – in other words, a matter of decency and shame, and nothing more! (Nietzsche 1974)

> Even the healthiest woman runs a zigzag course between sexual and individual life, stunting herself now as a person, now as a woman. (Andreas-Salomé 1978)

Woman's skepticism, Nietzsche suggests, comes from her disregard for truth. Truth does not concern her. Therefore, paradoxically, woman becomes the symbol of Truth, of that which constantly eludes man and must be won, which lures and resists, mocks and seduces, and will not be captured. This skepticism, this truth of nontruth, is the "affirmative woman" Nietzsche loved and was, Derrida suggests. It is the philosophical position Nietzsche himself occupies and speaks from – a position which Derrida locates in the terms of a rhetoric, "between the 'enigma of this solution' and the 'solution of this enigma'" (Derrida 1976b: 51).[1] The place from where he speaks, the locus of this enunciation, is a constantly shifting place within discourse (philosophy), a rhetorical function and construct; and a construct which – call it *différance*, displacement, negativity, internal exclusion, or marginality – has become perhaps the foremost rhetorical trope of recent philosophical speculation. However, in speaking from that place, from the position of woman, Nietzsche need not "stunt" himself "now as a person, now as a woman," as his contemporary and sometime friend Lou Andreas-Salomé admittedly did.[2] The difference between them, if I may put it bluntly, is not *différance* but gender.

If Nietzsche and Derrida can occupy and speak from the position of woman, it is because that position is vacant and, what is more, cannot be claimed by women. To anticipate a point that will be elaborated later on, I simply want to suggest that while the question of woman for the male philosophers is a question of style (of discourse, language, writing – of philosophy), for Salomé, as in most presentday feminist thinking, it is a question of gender – of the social construction of "woman" and "man," and the semiotic production of subjectivity. And whereas both style and gender have much to do with rhetoric, the latter (as I use the term and will attempt to articulate it) has also much to do with history, practices, and the imbrication of meaning with experience; in other words, with the mutually constitutive effects in semiosis of what Peirce called the "outer world" of social reality and the "inner world" of subjectivity.

With this in mind, let me then step into the role of Nietzsche's older woman and cast my considerations on the semiotic production of gender between the rhetoric of violence and the violence of rhetoric.

The very notion of a "rhetoric of violence" presupposes that some order of language, some kind of discursive representation is at work not only in the concept "violence" but in the social practices of violence as well. The (semiotic) relation of the social to the discursive is thus posed from the start. But once that relation is instated, once a connection is assumed between violence and rhetoric, the two terms begin to slide and, soon enough, the connection will appear to be reversible. From the Foucauldian notion of a rhetoric of violence, an order of language which speaks violence – names certain behaviors and events as violent, but not others, and constructs objects and subjects of violence, and hence violence as a social fact – it is easy to slide into the reverse notion of a language which, itself, produces violence. But if violence is in language, before if not regardless of its concrete occurrences in the world, then there is also a violence of rhetoric, or what Derrida has called "the violence of the letter" (1976a: 101–40).

I will contend that both views of the relation between rhetoric and violence contain and indeed depend on the same representation of sexual difference, whether they assume the "fact" of gender or, like Derrida, deny it: and, further, that the representation of violence is inseparable from the notion of gender, even when the latter is explicitly "deconstructed" or, more exactly, indicated as "ideology." I contend, in short, that violence is en-gendered in representation.

Violence en-gendered

In reviewing the current scholarship on family violence, Wini Breines and Linda Gordon begin by saying: "Only a few decades ago, the term 'family violence' would have had no meaning: child abuse, wife beating, and incest would have been understood but not recognized as serious social problems"

(Breines and Gordon 1983: 490). In particular, while child abuse had been ''discovered'' as far back as the 1870s, but later lost visibility, social science research on wife beating (more often called ''spouse abuse'' or ''marital violence'') is altogether recent; and incest, although long labeled a crime, was thought to be rare and, in any event, not related to (family) violence. In other words, the concept of a form of violence institutionally inherent – if not quite institutionalized – in the family, did not exist as long as the expression ''family violence'' did not.

Breines and Gordon, a sociologist and a historian, are keenly aware of the semiotic, discursive dimension of the social. Thus, they go on to argue, if the great majority of scholarly studies still come short of a coherent understanding of family violence as a social problem, the reason is that, with the exception of feminist writers, clinicians, and a few male empirical researchers, the work in this area fails to analyze the terms of its own inquiry, especially terms such as family, power, and gender. For, Breines and Gordon maintain, violence between intimates must be seen in the wider context of social power relations; and gender is absolutely central to the family. In fact, we may add, it is as necessary to the constitution of the family as it is itself, in turn, forcefully constructed and inevitably reproduced by the family. Moreover, they continue, institutions like the medical and other ''helping professions'' (such as the police and the judiciary) are complicit, or at least congruent, with ''the social construction of battering.'' For example, a study by Stark, Flitcraft, and Frazier (1979) of how the emergency room of a city hospital treated women for injuries or symptoms while completely ignoring the causes, if the injuries resulted from battering, shows how the institution of medicine ''coerce[s] women who are appealing for help back into the situations and relationships that batter them. It shows a system taking women who were hit, and turning them into battered women'' (Breines and Gordon 1983: 519).

The similarity of this critical position with that of Michel Foucault, himself a social historian, is striking, though no reference is made to his works (among them, *Discipline and Punish* and *The History of Sexuality* would be quite germane). But what the similarity makes apparent and even more striking is the difference of the two positions; that difference being, again, gender – not only the notion of gender, which is pivotal to the argument of Breines and Gordon, and largely irrelevant to Foucault's, but also, I will dare say, the gender of the authors. For it is feminism, the historical practice of the women's movement and the discourses which have emerged from it – like the collective speaking, confrontation, and reconceptualization of the female's experience of sexuality – that inform the epistemological perspective of Breines and Gordon. They refute the idea that all violence is of similar origin, whether that origin be located in the individual (deviance) or in an abstract, transhistorical notion of society (''a sick society''). And they counter the dominant representation of violence as a ''breakdown in social order'' by proposing instead that violence is a sign of ''a power struggle for the *maintenance* of a certain kind of social order'' (1983: 511). But which kind of social order is in question, to be maintained or to be dismantled, is just what is at stake in the discourse on family violence. It is also where Breines and Gordon differ from Foucault.

As they see it, both the intrafamily and the gender-neutral methodological perspectives on incest, for instance, which are often found combined, are motivated by the desire to explain away a reality too uncomfortable or threatening to nonfeminists. (In spite of the agreement among statistical studies that, in cases of incest as well as child sexual abuse, 92 per cent of the victims are females and 97 per cent of the assailants are males, ''predictably enough, until very recently the clinical literature ignored this feature of incest, implying that, for example, mother–son incest was as prevalent as father–daughter incest'' (1983: 523).) Such studies not only obscure the actual history of violence against women but, by disregarding the feminist critique of patriarchy, they effectively discourage analysis of family violence from a context of both societal and *male* supremacy. Following up on the insights provided by Breines and Gordon, one can see that this is undoubtedly the rhetorical function of gender-neutral expressions such as ''spouse abuse'' or ''marital violence,'' which at once imply that both spouses may equally engage in battering the other, and subtly hint at the writer's or speaker's nonpartisan stance of scientific and moral neutrality. In other words, even as those studies purport to remain innocent of the ideology or of the rhetoric of violence, they cannot avoid and indeed purposefully engage in the violence of rhetoric.

Foucault, for his part, is well aware of the paradox. The social, as he envisions it, is a field of forces, a crisscrossing of practices and discourses involving relations of power. With regard to the latter, individuals, groups, or classes assume variable positions, exercising at once power and resistance in an interplay of non-egalitarian but mobile, changeable relations; for the very existence of power relations ''depends on a multiplicity of points of resistance . . . present everywhere in the power network'' (Foucault 1980: 94). Both power and resistance, then, operate concurrently in ''the strategic field'' that constitutes the social, and both traverse or spread across – rather than inhere in or belong to – institutions, social stratifications, and individual unities. However, it is power, not resistance or negativity, that is the positive condition of knowledge. Far from being an agency of repression, power is a productive force that weaves through the social body as a network of discourses and generates simultaneously forms of knowledge and forms of subjectivity or what we call social subjects. Here, one would think, the rhetoric of power and the power of rhetoric are one and the same thing. Indeed, he writes,

> this history of sexuality, or rather this series of studies concerning the historical relationship of power and the discourse on sex is, I realize, a

> circular project in the sense that it involves two endeavors that refer back to one another. We shall try to rid ourselves of a juridical and negative representation of power, and cease to conceive of it in terms of law, prohibition, liberty, and sovereignty. But how then do we analyze what has occurred in recent history with regard to this thing – seemingly one of the most forbidden areas of our lives and bodies – that is sex? How, if not by way of prohibition and blockage, does power gain access to it? (1980: 90)

His answer posits the notion of a ''technology'' of sex, a set of ''techniques for maximizing life'' (1980: 123) developed and deployed by the bourgeoisie since the end of the eighteenth century in order to ensure its class survival and continued hegemony. Those techniques involved the elaboration of discourses (classification, measurements, evaluations, etc.) about four privileged ''figures'' or objects of knowledge: the sexualization of children and the female body, the control of procreation, and the psychiatrization of anomalous sexual behavior as perversion. These discourses – which were implemented through pedagogy, medicine, demography, and economics, were anchored or supported by the institutions of the state and became especially focused on the family – served to disseminate and to ''implant'' those figures and modes of knowledge into each individual, family, and institution. This technology ''made sex not only a secular concern but a concern of the state as well; to be more exact, sex became a matter that required the social body as a whole, and virtually all of its individuals, to place themselves under surveillance'' (1980: 116).

Sexuality, then, is not a property of bodies or something originally existent in human beings, but the product of that technology. What we call sexuality, Foucault states, is ''the set of effects produced in bodies, behaviors, and social relations'' by the deployment of ''a complex political technology'' (1980: 127), which is to say, by the deployment of sexuality. The analysis is in fact circular, however attractive or fitting. Sexuality is produced discursively (institutionally) by power, and power is produced institutionally (discursively) by the deployment of sexuality. Such a representation, like Foucault's view of the social, leaves no event or phenomenon out of the reach of *its* discursive power; nothing escapes from the discourse of power, nothing exceeds the totalizing power of discourse. His conclusion, therefore, is at best paradoxical. ''We must not think that by saying yes to sex, one says no to power. . . . The rallying point for the counterattack against the deployment of sexuality ought not to be sex-desire, but bodies and pleasures'' (1980: 157) – as if bodies and pleasures existed apart from the discursive order, from language or representation. But then they would exist in a space which his theory precisely locates outside the social.

I have suggested elsewhere that there may be a discrepancy between Foucault's theory and radical politics (his interventions in issues of capital punishment, prison revolts, psychiatric clinics, judiciary scandals, etc.), a discrepancy which can be accounted for by a contradiction perhaps inescapable at this time in history: the twin and opposite pull exerted on any progressive or radical thinker by the positivity of political action, on one front, and the negativity of critical theory, on the other. The contradiction is most evident, for me, in the efforts to elaborate a feminist theory of culture, history, representation, or subjectivity. Since feminism begins at home, so to speak, as a collective reflection on practice, on experience, on the personal as political, and on the politics of subjectivity, a feminist theory only exists as such in so far as it refers and constantly comes back to these issues. The contradictory pressure toward affirmative political action (the "counterattack") and toward the theoretical negation of patriarchal culture and social relations is glaring, unavoidable, and probably even constitutive of the specificity of feminist thought. In Foucault, the effect of that discrepancy (if my hypothesis is correct) has prompted charges of "paradoxical conservatism."[3]

For example, his political stance on the issue of rape, in the context of the reform of criminal law in France, has been criticized by French feminists as more subtly pernicious than the traditional "naturalist" ideology. Arguing for decriminalization (and the desexualization) of rape, in a volume published in 1977 by the Change collective with the title *La folie encerclée*, Foucault proposed that rape should be treated as an act of violence like any other, an act of aggression rather than as a sexual act. A similar position was also held by some American feminists (e.g., Brownmiller 1975), though with the opposite intent with regard to its juridical implications, and has been acutely criticized within American feminism: "taking rape from the realm of 'the sexual', placing it in the realm of 'the violent', allows one to be against it without raising any questions about the extent to which the institution of heterosexuality has defined force as a normal part of [(hetero)sexual relations]" (MacKinnon 1979: 219). In the terms of Foucault's theoretical analysis, his proposal may be understood as an effort to counter the technology of sex by breaking the bond between sexuality and crime; an effort to enfranchise sexual behaviors from legal punishment, and so to render the sexual sphere free from interventions by the state. Such a form of "local resistance" on behalf of the men imprisoned on, or subject to, charges of rape, however, would paradoxically but practically work to increase and further to legitimate the *sexual* oppression of women. As Monique Plaza put it, it is a matter of "our costs and their benefits." For what is rape if not a sexual practice, she asks, an act of *sexual* violence? While it may not be exclusively practiced on women, "rape is sexual essentially because it rests on the very social difference between the sexes. . . . It is *social sexing* which is latent in rape. If men rape women, it is precisely because they are women in a social sense"; and when a male is raped, he too is raped "as a woman" (Plaza 1980: 31).

This allows us to unravel the contradiction at the heart of Foucault's modest proposal, a contradiction which his analysis of sexuality does not serve to resolve: to speak against sexual penalization and repression, in our society, is to uphold the sexual oppression of women or, better, to uphold the practices and institutions that produce "woman" in terms of the sexual, and then oppression in terms of gender. (Which of course is not to say that oppression is not also produced in other terms.) To release "bodies and pleasures" from the legal control of the state, and from the relations of power exercised through the technology of sex, is to affirm and perpetuate the present social relations which give men rights over women's bodies. To decriminalize rape is, as Plaza states – making full use of the rhetoric of violence in her political confrontation with Foucault – to "defend the rights of the rapists . . . from the position of potential rapist that you are 'subjected' to by your status as a man" (1980: 33). Here Plaza sharply identifies the problem in Foucault's own "enunciative modality" (defined in Foucault 1972); that is to say, the place or sociosexual position from which he speaks, that of the male or male-sexed subject. For sexuality, not only in the general and traditional discourse, but in Foucault's as well, is not construed as gendered (as having a male form and a female form), but simply as male. Even when it is located, as it very often is, *in* the woman's body, sexuality is an attribute or property of the male. It is in this sense, in light of that "enunciative modality" common to all the accepted discourses in western culture (but not only there), that Adrienne Rich's notion of "compulsory heterosexuality" acquires its profoundest resonance and productivity. And in this sense her argument is not at the margins of feminism, as she seems to fear, but quite central to it (Rich 1980).

The historical fact of gender, the fact that it exists in social reality, that it has concrete existence in cultural forms and actual weight in social relations, makes gender a political issue that cannot be evaded or wished away, much as one would want to, be one male or female. For even as we agree that sexuality is socially constructed and overdetermined, we cannot deny the particular specification of gender that is the issue of that process; nor can we deny that precisely such a process finally positions women and men in an antagonistic and asymmetrical relation. The interests of men and women or, in the case in question, of rapists and their victims, are exactly opposed in the practices of social reality, and cannot be reconciled rhetorically. This is the blind spot in Foucault's radical politics and antihumanist theory, both of which must and do appeal to feminists as valuable contributions to the critique of ideology (see for example Martin 1982, and Doane, Mellencamp, and Williams 1984). Therefore, illuminating as his work is to our understanding of the mechanics of power in social relations, its critical value is limited by his unconcern for what, after him, we might call "the technology of gender" – the techniques and discursive strategies by which gender is constructed and hence, as I argue, violence is en-gendered.

But there may be another chestnut in the fire, another point at issue. To say that (a) the concept of "family violence" did not exist before the expression came into being, as I said earlier, is not the same as saying that (b) family violence did not exist before "family violence" became part of the discourse of social science. The enormously complex relation binding expression, content, and referent (or sign, meaning, and object) is what makes (a) and (b) not the same. It seems to me that of the three – the concept, the expression, and the violence – only the first two belong to Foucault's discursive order. The third is somewhere else, like "bodies and pleasures," outside the social. Now, for those of us whose bodies and whose pleasures are out there where the violence is (in that we have no language, enunciative position, or power apparatuses to speak them), the risk of saying yes to sex-desire and power is relatively small, and amounts to a choice between the devil and the deep blue sea. If we then want to bring our bodies and our pleasures closer, where we might see what they are like; better still, where we might represent them from another perspective, construct them with another standard of measurement, or understand them within other terms of analysis; in short, if we want to attempt to know them, we have to leave Foucault and turn, for the time being, to Peirce.

For Peirce, the object has more weight, as it were. The real, the physical world and empirical reality, are of greater consequence to the human activity of semiosis, as outlined by Charles Sanders Peirce, than they are to the symbolic activity of signification, as defined in Saussure's theory of language and re-elaborated in contemporary French thought. Saussure's insistence on the arbitrary or non-motivated nature of the linguistic sign caused semiology to extend the categorical distinction between language (*langue*, the language system) and reality to all forms and processes of representation, and thus to posit an essential discontinuity between the orders of the symbolic and the real. Thereafter, not only would the consideration of the referent be no longer pertinent – or even possible – to the account of signification processes; but the different status of the signifier and the signified would be questioned. The signified would be seen as either inaccessible, separated from the signifier by the "bar" of repression (Lacan 1966: 497), or equally engaged in the "play of differences" that make up the system of signifiers and the domain of signification (Derrida 1976a: 7). The work of the sign, in brief, would have no reference and no purchase on the real. For Peirce, on the other hand, the "outer world" enters into semiosis at both ends of the signifying process: first through the object, more specifically the "dynamic object," and secondly through the final interpretant. This complicates the picture in which a signifier would immediately correspond to a signified (Saussure) or merely refer to another signifier (Lacan, Derrida). Take the famous definition:

> A sign, or representamen, is something which stands to somebody for something in some respect or capacity. It addresses somebody, that is, it creates in the mind of that person an equivalent sign, or perhaps a more developed sign. That sign which it creates I call the *interpretant* of the first sign. The sign stands for something, its *object*. It stands for that object, not in all respects, but in reference to a sort of idea, which I have sometimes called the *ground* of the representation. (Peirce 2: 228)

As Umberto Eco observes in his brilliant essay on "Peirce and the Semiotic Foundations of Openness" (Eco 1979: 175–99), the notions of meaning, ground, and interpretant all pertain in some degree to the area of the signified, while interpretant and ground also pertain to some degree to the area of the referent (object). Moreover Peirce distinguishes between the dynamic object and the immediate object, and it is the notion of ground that sustains the distinction. The dynamic object is external to the sign: it is that which "by some means contrives to determine the sign to its representation" (4: 536). The immediate object, instead, is internal; it is an "Idea" or a "mental representation," "the object as the sign itself represents it" (4: 536).

From the analysis of the notion of "ground" (a sort of context of the sign, which makes pertinent certain attributes or aspects of the object and thus is already a component of meaning), Eco argues that not only does the sign in Peirce appear as a textual matrix; the object, too, "is not necessarily a thing or a state of the world but a rule, a law, a prescription: it appears as the operational description of a set of possible experiences" (Eco 1979: 181).

> Signs have a direct connection with Dynamic Objects only insofar as objects determine the formation of a sign; on the other hand, signs only "know" Immediate Objects, that is, meanings. There is a difference between the *object of which a sign is a sign* and the *object of a sign*: the former is the Dynamic Object, a state of the outer world; the latter is a semiotic construction. (Eco 1979: 193)

But the immediate object's relation to the representamen is established by the interpretant, which is itself another sign, "perhaps a more developed sign." Thus, in the process of unlimited semiosis, the nexus object–sign–meaning is a series of ongoing mediations between "outer world" and "inner" or mental representations. The key term, the principle that supports the series of mediations, is of course the interpretant.

As Peirce sees it, "the problem of what the 'meaning' of an intellectual concept is can only be solved by the study of the interpretants, or proper significate effects, of signs (5: 475). He then describes three general classes.

1. "The first proper significate effect of a sign is a *feeling* produced by it." This is the *emotional* interpretant. Although its "foundation of truth"

may be slight at times, often this remains the only effect produced by a sign such as, for example, the performance of a piece of music.

2. When a further significate effect is produced, however, it is "through the mediation of the emotional interpretant"; and this second type of meaning effect he calls the *energetic* interpretant, for it involves an "effort," which may be a muscular exertion but is more usually a mental effort, "an exertion upon the Inner World."

3. The third and final type of meaning effect that may be produced by the sign, through the mediation of the former two, is "a *habit-change*": "a modification of a person's tendencies toward action, resulting from previous experiences or from previous exertions." This is the "ultimate" interpretant of the sign, the effect of meaning on which the process of semiosis, in the instance considered, comes to rest. "The real and living logical conclusion *is* that habit," Peirce states, and designates the third type of significate effect, the *logical* interpretant. But immediately he adds a qualification, distinguishing this logical interpretant from the concept or "intellectual" sign:

> The concept which is a logical interpretant is only imperfectly so. It somewhat partakes of the nature of a verbal definition, and is as inferior to the habit, and much in the same way, as a verbal definition is inferior to the real definition. The deliberately formed, self-analyzing habit – self-analyzing because formed by the aid of analysis of the exercises that nourished it – is the living definition, the veritable and final logical interpretant. (5: 491)

The final interpretant, then, is not "logical" in the sense in which a syllogism is logical, or because it is the result of an "intellectual" operation like deductive reasoning. It is logical in that it is "self-analyzing," or, we might say, in that it makes sense of the emotion and muscular/mental effort which preceded it by providing a conceptual representation of that effort. Such a representation is implicit in the notion of habit as a "tendency toward action" and in the solidarity of habit and belief (5: 538).

Peirce's formulation of the ultimate interpretant maps another path or a way back from semiosis to reality. For Eco, it provides the "missing link" between signification and concrete action. The final interpretant, he states, is not a Platonic essence or a transcendental law of signification but a result, as well as a rule: "to have understood the sign as a rule through the series of its interpretants means to have acquired the habit to act according to the prescription given by the sign. . . . The action is the place in which the *haeccéitas* ends the game of semiosis" (Eco 1979: 194–5). But we should go further in our reading of Peirce, and so enter into a territory where Eco fears to tread, the terrain of subjectivity.

When Peirce speaks of habit as the result of a process involving emotion, muscular and mental exertion, and some kind of conceptual representation (the "final logical interpretant"), he is thinking of individual persons as the subject of such a process. If the modification of consciousness, the habit or habit-change, is indeed the meaning effect, the "real and living" conclusion of each single process of semiosis, then where "the game of semiosis" ends, time and time again, is not exactly "concrete action," as Eco sees it, but a person's (subjective) disposition, a readiness (to action), a set of expectations. For the chain of meaning comes to a halt, however temporarily, by anchoring itself to somebody, to some body, an individual subject.[4] Thus, as we use signs or produce interpretants, their significant effects must pass through each of us, each body and each consciousness, before they may produce an effect or an action upon the world. Finally, then, the individual's habit as a semiotic production is both the result and the condition of the social production of meaning.

Clearly, this reading of Peirce points toward a possible elaboration of semiotics as a theory of culture that hinges on a historical, materialist, *and* gendered subject – a project that cannot be pursued here. What I wish to stress, for the sake of the present discussion, is the sense of a certain weight of the object in semiosis, an overdetermination wrought into the work of the sign by the real, or what we take as reality, even if it is itself already an interpretant; and hence the sense that experience (habit), however misrecognized or misconstrued, is indissociable from meaning; and therefore that practices – events and behaviors occurring in social formations – weigh in the constitution of subjectivity as much as does language. In that sense, too, violence is not simply "in" language or "in" representation, but it is also thereby en-gendered.

Violence and representation

When one first surveys the representations of violence in general terms, there seem to be two kinds of violence with respect to its object: male and female. I do not mean by this that the "victims" of such kinds of violence are men and women, but rather that the object on which or to which the violence is done is what establishes the meaning of the represented act; and that object is perceived or apprehended as either feminine or masculine. An obvious example of the first instance is "nature," as in the expression "the rape of nature," which at once defines nature as feminine, and rape as violence done to a feminine other (whether its physical object be a woman, a man, or an inanimate object). Speculating on the particular rhetoric of violence that permeates the discourse in which scientists describe their encounter with the unknown, Evelyn Fox Keller finds a recurrent thematics of conquest, domination, and aggression reflecting a "basic adversarial relation to the object of study."

Problems, for many scientists are things to be "attacked," "licked" or

> "conquered." If more subtle means fail, then one resorts to "brute force," to the "hammer and tongs" approach. In the effort to "master" nature, to "storm her strongholds and castles," science can come to sound like a battlefield. Sometimes, such imagery becomes quite extreme, exceeding even the conventional imagery of war. Note, for example, the language in which one scientist describes his pursuit: "I liked to follow the workings of another mind through these minute, teasing investigations to see a relentless observer get hold of Nature and squeeze her until the sweat broke out all over her and her sphincters loosened." (Keller 1983: 20)

The "genderization of science," as Keller calls the association of scientific thought with masculinity and of the scientific domain with femininity, is a pervasive metaphor in the discourse of science, from Bacon's prescription of a "chaste and lawful marriage between Mind and Nature" to Bohr's chosen emblem, the yin-yang symbol, for his coat of arms (Keller 1978: 413, 423). It is a compelling representation, whose effects for the ideology and the practice of science, as well as for the subjectivity of individual scientists, are all the more forceful since the representation is treated as a myth; that is to say, while the genderization of science is admitted and encouraged in the realm of common knowledge, it is simultaneously denied entry or currency in the realm of formal knowledge (Keller 1978: 410). This is the case not only in the "hard" sciences, so-called, but also more often than not in the "softer" disciplines and even, ironically enough, in the study of myth.

The other kind of violence is that which in *Violence and the Sacred* René Girard has aptly called "violent reciprocity," the acting out of "rivalry" between brothers or between father and son, and which is socially held in check by the institution of kinship, ritual, and other forms of mimetic violence (war and sport come immediately to mind). The distinctive trait here is the "reciprocity" and thus, by implication, the equality of the two terms of the violent exchange, the "subject" and the "object" engaged in the rivalry; and consequently the masculinity attributed, in this particular case, to the object. For the subject of the violence is always, by definition, masculine; "man" is by definition the subject of culture and of any social act.[5]

In the mythical text, for example, according to Lotman's theory of plot typology, there are only two characters, the hero and the obstacle or boundary. The first is the mythical subject, who moves through the plot-space establishing differences and norms. The second is but a function of that space, a marker of boundary, and therefore inanimate even when anthropomorphized.

> Characters can be divided into those who are mobile, who enjoy freedom with regard to plot-space, who can change their place in the structure of the artistic world and cross the frontier, the basic topological feature of this space, and those who are immobile, who represent, in fact, a function of

> this space. Looked at typologically, the initial situation is that a certain plot-space is divided by a *single* boundary into an internal and an external sphere, and a *single* character has the opportunity to cross that boundary. . . . Inasmuch as closed space can be interpreted as "a cave," "the grave", "a house," "woman" (and, correspondingly, be allotted the features of darkness, warmth, dampness), entry into it is interpreted on various levels as "death," "conception," "return home" and so on; moreover all these acts are thought of as mutually identical. (Lotman 1979: 167–8)

In the mythical text, then, the hero must be male regardless of the gender of the character, because the obstacle, whatever its personification (sphinx or dragon, sorceress or villain), is morphologically female – and, indeed, simply the womb, the earth, the space of his movement. As he crosses the boundary and "penetrates" the other space, the mythical subject is constructed as human being and as male; he is the active principle of culture, the establisher of distinction, the creator of differences. Female is what is not susceptible to transformation, to life or death, she (it) is an element of plot-space, a topos, a resistance, matrix, and matter.

Narrative cinema, too, performs a similar inscription of gender in its visual figuration of the masculine and the feminine positions. The woman, fixed in the position of icon, spectacle or image to be looked at, bears the mobile look of both the spectator and the male character(s). It is the latter who commands at once the action and the landscape, and who occupies the position of subject of vision, which he relays to the spectator. As Laura Mulvey shows in her analysis of the complex relations of narrative and visual pleasure, "sadism demands a story" (1975: 14). Thus, if Oedipus has become a paradigm of human life and error, narrative temporality and dramatic structure, one may be entitled to wonder whether that is purely due to the artistry of Sophocles or the widespread influence of Freud's theory of human psychic development in our culture; or whether it might not also be due to the fact that, like the best of stories and better than most, the story of Oedipus weaves the inscription of violence (and family violence, at that) into the representation of gender.

I will now turn to two celebrated critical texts, which exemplify two discursive strategies deployed in the construction of gender and two distinctive rhetorical configurations of violence. The first is Lévi-Strauss's reading in "The Effectiveness of Symbols" (Lévi-Strauss 1967), of a Cuna incantation performed to facilitate difficult childbirth; a reading which prompts him to make a daring parallel between shamanistic practices and psychoanalysis, and allows him to elaborate his crucial notion of the unconscious as symbolic function. The shaman's cure consists, he states, "in making explicit a situation originally existing on the emotional level and in rendering acceptable to the mind pains which the body refuses to tolerate" by provoking an experience "through symbols, that is, through meaningful equivalents of things meant which belong to another order of reality" (1967: 192, 196). Whereas the arbitrary pains are alien and unacceptable to the woman, the supernatural monsters evoked by the shaman in his symbolic narrative are part of a coherent system on which the native conception of the universe is founded. By calling upon the myth, the shaman reintegrates the pains within a conceptual and meaningful whole, and "provides the sick [*sic*] woman with a *language*, by means of which unexpressed, and otherwise inexpressible, psychic states can be immediately expressed" (1967: 193). Both the shaman's cure and psychoanalytic therapy, argues Lévi-Strauss, albeit with an inversion of all the elements, are done by means of a manipulation carried out through symbols that constitute a meaningful code, a language.

Let us consider now the structure of the myth in question and the performative value of the shaman's narrative. For, after all, the incantation is a ritual, though based on myth. It has, that is, a practical purpose: it seeks to effect a physical, somatic transformation in its addressee. The main actors are the shaman, performing the incantation, and the woman in labor whose body is to undergo the transformation, to become actively engaged in expelling the full-grown fetus and bringing forth the child. In the myth which subtends the incantation, one would think, the hero must be a woman or at least a female spirit, goddess, or totemic ancestor. But it is not so. Not only is the hero a male, personified by the shaman, as are his helpers, also symbolized with decidedly phallic attributes; and not only is the incantation intended to effect the childbearing woman's identification with the male hero in his struggle with the villain (a *female* deity who has taken possession of the woman's body and soul). But, more importantly, the incantation aims at detaching the woman's identification or perception of self from her own body. It seeks to sever her identification with a body which she must come to perceive precisely as a space, the territory in which the battle is waged. The hero's victory then results in his recapturing the woman's soul, and his descent through the landscape of her body symbolizes the (now) unimpeded descent of the fetus along the birth canal.

The effectiveness of symbols, the work of the symbolic function in the unconscious, would thus effect a splitting of the female subject's identification into the two mythical positions of hero (the human subject) and boundary (spatially fixed object or personified obstacle – her body). The doubt that the apprehension of one's body or oneself as obstacle, landscape, or battlefield may not "provide the . . . woman with a language" does not cross the text. But whether or not this construct would "make sense" to the Cuna woman for whose benefit the ritual is presumably performed, Lévi-Strauss's interpretation must be acceptable in principle to Lotman, Girard, and any others who look on the history of the human race from the anthropological perspective and within an epistemology wherein "biological" sexual difference is the ground (in Peirce's term) of gender. In that perspective, woman remains outside of history. She is mother and nature, matrix and matter, "an equivalent more universal than money," as Lea Melandri accurately phrased it (1977: 27). The discourse of the sciences of man constructs the object as female and the female as object. This, I suggest, is its rhetoric of violence, even when the discourse presents itself as humanistic, benevolent or well-intentioned.

Indeed Derrida criticizes Lévi-Strauss's paternalistic attitude toward his objects of study (the Nambikwara), as well as the naïvety by which he regards them as an "innocent" people because they have no written language. In such a community, described in the autobiographical *Tristes Tropiques*, violence would be introduced by western civilization, and actually erupts as the anthropologist (Lévi-Strauss himself, who recounts the event) teaches a group of children how to write. The "revenge" of one little girl, struck by another during the "Writing Lesson," consists in revealing to the anthropologist the "secret" of the other girl's proper name, which the Nambikwara are not allowed to use. What is ingenuous, for Derrida, is Lévi-Strauss's ostensible belief that writing is merely the phonetic notation of speech, and that violence is an effect of written language (civilization) rather than of language as such; for "all societies capable of producing, that is to say of obliterating, their proper names, and of bringing classificatory difference into play, practice writing in general" (Derrida 1976a: 109).

> To name, to give names that it will on occasion be forbidden to pronounce, such is the originary violence of language which consists in inscribing within a difference, in classifying, in suspending the vocative absolute. To think the unique *within* the system, to inscribe it there, such is the gesture of the arche-writing: arche-violence, loss of the proper, of absolute proximity, of self-presence. . . . Out of this arche-violence, forbidden and therefore confirmed by a second violence that is reparatory, protective, instituting the "moral," prescribing the concealment of writing and the effacement and obliteration of the so-called proper name which was already dividing the proper, a third violence can *possibly* emerge or not (an empirical possibility) within what is commonly called evil, war, indiscretion, rape: which consists of revealing by effraction the so-called proper name, the originary violence which has severed the proper from its property and its self-sameness [*propreté*]. (1976a: 112)

Empirical or common violence (and we cannot help remarking the text's own classificatory play in the listing of signifiers: evil, war, indiscretion, rape) is "more complex" than the other two levels to which it refers, namely, arche-violence and law. Unfortunately for us, however, Derrida is not concerned to analyze it or to suggest why, how, or when it may possibly emerge. He only implies that the emergence of empirical violence, the fact of violence in society, is no accident, though Lévi-Strauss would need to see it as an accident in order to maintain his belief in the natural innocence and goodness of the primitive culture. From Rousseau and the eighteenth century, Derrida concludes, Lévi-Strauss has inherited an archeology which "is also a teleology and an eschatology": "The dream of a full and immediate presence closing history [suppresses] contradiction and difference" (1976a: 115).

The rhetorical construct of a "violence of the letter," the originary violence which pre-empts presence, identity, and property or propriety, is perhaps more accessible in another of Derrida's own works, *Spurs*, where he performs a reading of Nietzsche and, with him, addresses just what he claimed that Lévi-Strauss suppressed – contradiction and difference. This could be my second textual *exemplum*, whereby to illustrate what I earlier called the violence of rhetoric. It would support my contention that, while Derrida's discourse denies the fact of gender, its "becoming woman" depends on the same construct of sexual difference precisely if naively and traditionally articulated by Lévi-Strauss (1969).

Were I to do so, however, I would earn Derrida's contempt for "those women feminists so derided by Nietzsche," I would put myself in the position of one "who aspires to be like a man," who "seeks to castrate" and "wants a castrated woman" (Derrida 1976b: 53). I shall not do so, therefore. Decency and shame prevent me, though nothing more. I shall instead approach Derrida's text obliquely – a gesture the philosopher may not find displeasing – by way of another's reading, or a quadruple displacement, if you will.

"The discourse of man," writes Gayatri Spivak, "is in the metaphor of woman" (1983: 169). The problem with phallocentrism "is not merely one of psycho-socio-sexual behavior [as, we recall, Foucault would have it] but of the production and consolidation of reference and meaning" (1983: 169). Derrida's critique of phallocentrism – deconstruction – takes the woman as "model" for the deconstructive discourse. It takes the woman as model because, as Spivak reads (Derrida reading) Nietzsche, the woman can fake an orgasm, while the man cannot: "Women impersonate themselves as having an orgasm even at the time of orgasm. Within the historical understanding of women as incapable of orgasm, Nietzsche is arguing that impersonating is woman's only sexual pleasure" (Spivak 1983: 170). Thus, in what appears to me as a case of inscribing gender with a vengeance, Derrida searches for the name of the mother in *Glas*; elsewhere, he uses the "name of woman" to question the "we-men" of the philosophers (1983: 173); and *Dissemination* takes the hymen as figure for the text, the undecidability of meaning, the "law of the textual operation – of reading, writing, philosophizing" (1983: 175).

Deconstruction thus effects "a feminization of the practice of philosophy," Spivak observes (with a phrase that reminds me immediately of Keller's "genderization of science"), adding that she does not regard it as "just another example of the masculine use of woman as instrument of self-assertion" (1983: 173). For if man can never "fully disown his status as subject,'' and if desire must still ''be expressed as man's desire,'' yet the deconstructor's enterprise – seeking his own displacement ''by taking the woman as object or figure'' – is an ''unusual and courageous'' one. Regretfully, one must infer, Spivak is led to admit that the question of woman, asked in the way Nietzsche and Derrida ask it, ''is *their* question, not *ours*'' (1983: 184). Then she suggests, ''with respect,'' that such a feminization of philosophy as serves the male deconstructor ''might find its most adequate legend in male homosexuality defined as criminality, and that it cannot speak for the woman'' (1983: 177). One can only conclude that, in so far as the ''deconstructor'' is a woman, the value of that critical practice (''the 'patriarchy's' own self-critique'') is at best ambiguous. We can produce, as Spivak recommends, ''useful and scrupulous fake readings in the place of the passively active fake orgasm'' (1983: 186), but we will not have come at all closer to understanding, representing, or reconstructing our bodies and our pleasures otherwise.

For the female subject, finally, gender marks the limit of deconstruction, the rocky bed (so to speak) of the ''abyss of meaning.'' Which is not to say that woman, femininity, or femaleness are any more or any less outside discourse than anything else is. This is precisely the insistent emphasis of feminist criticism: gender must be accounted for. It must be understood not as a ''biological'' difference that lies before or beyond signification, nor as a culturally constructed object of masculine desire, but as semiotic difference – a different production of reference and meaning such as, not Derrida and not Foucault, but possibly Peirce's notion of semiosis may allow us to begin to chart. Clearly, the time of ''replacing feminist criticism'' (Kamuf 1982) has not come.

Notes

1 In Barbara Harlow's translation of *Spurs*, the quotations from Nietzsche incorporated in Derrida's text are given in the words of the English translation by Thomas Common (*Joyful Wisdom*, New York: Frederick Ungar, 1960). I have preferred to use Walter Kaufmann's translation in *The Gay Science* (1974), both below and, somewhat modified, in my epigraph above, which is from paragraph 64. In the passage cited by Derrida from *Die fröhliche Wissenschaft* (§ 71, ''On Female Chastity''), Nietzsche is speaking of the contradiction which upper-class women, reared in total ignorance of sexuality, must encounter at the moment of marriage. From their supposed ignorance of sex, Nietzsche mockingly laments, women are ''hurled, as by a gruesome lightning bolt, into reality and knowledge, by marriage – precisely by the man they love and esteem most! To catch love and shame in contradiction and to be forced to experience at the same time delight, surrender, duty, pity, terror, and who knows what else, in the face of the unexpected neighborliness of god and beast! . . . Even the compassionate curiosity of the wisest student of humanity is inadequate for guessing how this or that woman manages to accommodate herself to *this solution of the riddle*, and to *the riddle of a solution*, and what dreadful, far-reaching suspicions must stir in her poor, unhinged soul – and how the ultimate philosophy and skepsis of woman casts

anchor at this point!'' I have italicized the phrases which Derrida takes out of context and recalls in the frame of his interpretation of Nietzsche. As will be discussed later, Derrida reads in Nietzsche a progressive valorization of woman as a self-affirming power, ''a dissimulatress, an artist, a dionysiac''; and this is the ''affirmative woman'' that Derrida takes as his model for ''writing,'' for the critical operation of questioning, doubting, or ''deconstructing'' all truths.

2 For an interesting discussion of Andreas-Salomé's writing, figure, and historiographical ''legend'' from the perspective of present-day feminism, see Martin (1982). The quotation from Andreas-Salomé's *Zur Psychologie der Frau*, which appears at the beginning of this essay, is cited in Martin (1982: 29).

3 ''Paradoxical conservatism,'' I have argued,

> is a very appropriate phrase for a major theoretician of social history who writes of power and resistance, bodies and pleasures and sexuality as if the ideological structures and effects of patriarchy had nothing to do with history, as if they had no discursive status or political implications. The rape and sexual extortion performed on little girls by young and adult males is a ''bit of theater,'' a petty ''everyday occurrence in the life of village sexuality,'' purely ''inconsequential bucolic pleasures'' [Foucault 1980: 31–2]. What really matters to the historian is the power of institutions, the mechanisms by which these bits of theater become, he claims, pleasurable for the individuals involved – the men *and* the women, former little girls – who thus become complicit with those institutional apparati (de Lauretis 1984: 94).

This passage, which I take the liberty of reprinting here, occurs in the context of my analysis of a film, Nicholas Roeg's *Bad Timing: A Sensual Obsession* (1980), in light of some of Foucault's ideas. The film is an interesting study of ''marital violence,'' and an excellent visual-narrative text for a discussion of violence, representation, and gender.

4 My reading of Peirce's definition of the sign, and thus of the relationship of sign and subject, bears a comparison with Lacan's ostensibly antithetical formula (''a signifier represents a subject for another signifier''). I must again refer interested readers to chapter 6 of my book (1984): ''Semiotics and Experience,'' where a fuller discussion of Eco is also to be found.

5 Studies in language usage demonstrate that, if the term ''man'' includes women (while the obverse is not true, for the term ''woman'' is always gendered, i.e., sexually connoted), it is only to the extent that, in the given context, women are (to be) perceived as non-gendered ''human beings,'' and thus as man (see Spender (1980)). For example, Lévi-Strauss's theory of kinship (1969) is based on the thesis that women are both like men and unlike men: they are human beings (like men), but their special function in culture and society is to be exchanged and circulated among men (unlike men). Because of their ''value'' as means of sexual gratification and reproduction, women are the means – objects and signs – of social communication (among human beings). Nevertheless, as he is unwilling to exclude women from humanity or ''mankind,'' he compromises by saying that women are also human beings, although in the symbolic order of culture they do not speak, desire, or produce meaning *for themselves*, as men do, by means of the exchange of women. One can only conclude that, in so far as women are human beings, they are (like) men.

References

Andreas-Salomé, Lou (1978) *Zur Psychologie der Frau*. Ed. Gisela Brinker-Gabler. Frankfurt: Fischer Taschenbuch Verlag.

Breines, Wini and Gordon, Linda (1983) "The New Scholarship on Family Violence." *Signs: A Journal of Women in Culture and Society* 8 (3): 490–531.

Brownmiller, Susan (1975) *Against Our Will: Men, Women and Rape*. New York: Simon & Schuster.

Change (1977) *La folie encerclée*. Paris: Seghers/Laffont.

Derrida, Jacques (1976a) *Of Grammatology*. Trans. Gayatri Chakravorty Spivak. Baltimore and London: Johns Hopkins University Press.

——— (1976b) *Éperons. Les styles de Nietzsche*. Venice: Corbo e Fiore. (This is a four-language edition; the English translation is by Barbara Harlow.)

Doane, Mary Ann, Mellencamp, Patricia, and Williams, Linda (eds) (1984) *Re-Vision: Essays in Feminist Film Criticism*. Los Angeles: American Film Institute.

Eco, Umberto (1979) *The Role of the Reader: Exploration in the Semiotics of Texts*. Bloomington and London: Indiana University Press.

Foucault, Michel (1972) *The Archaeology of Knowledge*. Trans. A.M. Sheridan Smith. London: Irvington.

——— (1980) *The History of Sexuality, Vol. I: An Introduction*. Trans. Robert Hurley. New York: Vintage.

Girard, René (1977) *Violence and the Sacred*. Trans. Patrick Gregory. Baltimore and London: Johns Hopkins University Press.

Kamuf, Peggy (1982) "Replacing Feminist Criticism." *Diacritics* 12: 42–7.

Keller, Evelyn Fox (1978) "Gender and Science." *Psychoanalysis and Contemporary Thought* September: 409–33.

——— (1983) "Feminism as an Analytic Tool for the Study of Science." *Academe* September–October: 15–21.

Lacan, Jacques (1966) *Écrits*. Paris: Seuil.

Lauretis, Teresa de (1984) *Alice Doesn't: Feminism, Semiotics, Cinema*. Bloomington: Indianna University Press.

Lévi-Strauss, Claude (1961) *Tristes Tropiques*. Trans. John Russell. New York: Criterion.

——— (1967) *Structural Anthropology*. Trans. Claire Jacobson and Brooke Grundfest Schoept. Garden City, NY: Anchor.

——— (1969) *The Elementary Structures of Kinship*. Trans. James Harle Bell, John Richard von Sturmer, and Rodney Needham. Boston: Beacon.

Lotman, Jurij (1979) "The Origin of Plot in the Light of Typology." Trans. Julian Graffy. *Poetics Today* 1(1–2): 161–84.

MacKinnon, Catharine (1979) *Sexual Harassment of Working Women: A Case of Sex Discrimination*. New Haven, Conn.: Yale University Press.

Martin, Biddy (1982) "Feminism, Criticism, and Foucault." *New German Critique* 27: 3–30.

Melandri, Lea (1977) *L'infamia originaria*. Milan: Edizioni L'Erba Voglio.

Mulvey, Laura (1975) "Visual Pleasure and Narrative Cinema." *Screen* 16(3): 6–18.

Nietzsche, Friedrich (1974) *The Gay Science*. Trans. by Walter Kaufmann. New York: Vintage.

Peirce, Charles Sanders (1931–58) *Collected Papers*. Ed. Charles Hartshorne and Paul Weiss. Cambridge, Mass.: Harvard University Press.

Plaza, Monique (1980) "Our Costs and their Benefits." Trans. Wendy Harrison. *m/f* 4: 28–39. Originally in *Questions féministes* 3 (May 1978).

Rich, Adrienne (1980) "Compulsory Heterosexuality and Lesbian Existence." *Signs: A Journal of Women in Culture and Society* 5(4): 631–60.

Spender, Dale (1980) *Man Made Language*. London: Routledge & Kegan Paul.

Spivak, Gayatri Chakravorty (1983) "Displacement and the Discourse of Woman," in Mark Krupnick (ed.) *Displacement: Derrida and After*. Bloomington: Indiana University Press: 169–95.

Stark, Evan, Flitcraft, Anne, and Frazier, William (1979) "Medicine and Patriarchal Violence: The Social Construction of a 'Private' Event." *International Journal of Health Services* 9(3): 461–93.

Index